Microelectromechanical Systems—
Materials and Devices IV

MATERIALS RESEARCH SOCIETY
SYMPOSIUM PROCEEDINGS VOLUME 1299

Microelectromechanical Systems—Materials and Devices IV

Symposium held November 29–December 3, Boston, Massachusetts, U.S.A.

EDITORS

Frank W. DelRio

National Institute of Standards and Technology
Gaithersburg, Maryland, U.S.A.

Maarten P. de Boer

Carnegie Mellon University
Pittsburgh, Pennsylvania, U.S.A.

Christoph Eberl

Karlsruhe Institute of Technology
Karlsruhe, Germany

Evgeni Gusev

Qualcomm MEMS Technologies
San Jose, California, U.S.A.

Materials Research Society
Warrendale, Pennsylvania

CAMBRIDGE
UNIVERSITY PRESS

CAMBRIDGE UNIVERSITY PRESS
Cambridge, New York, Melbourne, Madrid, Cape Town,
Singapore, São Paulo, Delhi, Mexico City

Cambridge University Press
32 Avenue of the Americas, New York NY 10013-2473, USA

Published in the United States of America by Cambridge University Press, New York

www.cambridge.org
Information on this title: www.cambridge.org/9781107406834

Materials Research Society
506 Keystone Drive, Warrendale, PA 15086
http://www.mrs.org

First published 2011
First paperback edition 2012

Single article reprints from this publication are available through
University Microfilms Inc., 300 North Zeeb Road, Ann Arbor, MI 48106

CODEN: MRSPDH

ISBN 978-1-605-11276-3 Hardback
ISBN 978-1-107-40683-4 Paperback

CONTENTS

MATERIAL DEVELOPMENT AND OPTIMIZATION

PROCESS INTEGRATION

MICRO- AND NANOSENSORS

MATERIAL AND DEVICE RELIABILITY

PREFACE

Symposium S, "Microelectromechanical Systems—Materials and Devices IV," held November 29–December 3 at the 2010 MRS Fall Meeting in Boston, Massachusetts, focused on micro- and nanoelectromechanical systems (MEMS/NEMS), technologies which were spawned from the fabrication and integration of small-scale mechanical, electrical, thermal, magnetic, fluidic, and optical sensors and actuators with micro-electronic components. MEMS and NEMS have enabled performance enhancements and manufacturing cost reductions in a number of applications, including optical displays, acceleration sensing, radio-frequency switching, drug delivery, chemical detection, and power generation and storage. Although originally based on silicon microelectronics, the reach of MEMS and NEMS has extended well beyond traditional engineering materials, and now includes nanomaterials (nanotubes, nanowires, nanoparticles), smart materials (piezoelectric and ferroelectric materials, shape memory alloys, pH-sensitive polymers), metamaterials, and biomaterials (ceramic, metallic, polymeric, composite-based implant materials). While these new materials provide more freedom with regards to the design space of MEMS and NEMS, they also introduce a number of new fabrication and characterization challenges not previously encountered with silicon-based technology.

The symposium was devoted to addressing these challenges by providing a common forum for materials researchers and device engineers to discuss the relationships between MEMS and NEMS materials and device design, fabrication, performance, and reliability. 32 papers from the symposium are included in this proceedings volume. Of these, nine deal with material development and optimization, seven are devoted to process integration, eight explore new micro- and nanosensors, and eight look to address various aspects of material and device reliability. Thus, the volume focuses on both newer materials in the development and integration stages and traditional materials in the device optimization and reliability stages. Furthermore, it is worthwhile to note the continued success of MEMS and NEMS in drug delivery (one paper), cell manipulation and analysis (one paper), and power generation and storage (one paper), as these topics are critical to the emerging areas of nanomedicine and renewable energy.

This volume represents the fourth installment in a series of proceedings by MRS on this topic; the first three volumes were published as volumes 1052, 1139, and 1222. The fifth symposium in this series is scheduled for the 2011 MRS Fall Meeting.

Frank W. DelRio
Maarten P. de Boer
Christoph Eberl
Evgeni Gusev

February 2011

MATERIALS RESEARCH SOCIETY SYMPOSIUM PROCEEDINGS

MATERIALS RESEARCH SOCIETY SYMPOSIUM PROCEEDINGS

Prior Materials Research Society Symposium Proceedings available by contacting Materials Research Society

Material Development and Optimization

Mater. Res. Soc. Symp. Proc. Vol. 1299 © 2011 Materials Research Society
DOI: 10.1557/opl.2011.377

Biodegradable Microfluidic Scaffolds with Tunable Degradation Properties from Amino Alcohol-based Poly(ester amide) Elastomers

Jane Wang[1,2,3], Tatiana Kniazeva[2], Carly F. Campbell[2], Robert Langer[3,4], Jeffrey S. Ustin[5], Jeffrey T. Borenstein[2]

[1]Department of Materials Science and Engineering, Massachusetts Institute of Technology, Cambridge, Massachusetts, USA, 02139
[2]Biomedical Engineering, Charles Stark Draper Laboratory, Cambridge, MA, USA, 02139
[3]Program of Polymer Science and Technology, Massachusetts Institute of Technology, Cambridge, Massachusetts, USA, 02139
[4]Department of Chemical Engineering, Massachusetts Institute of Technology, Cambridge, Massachusetts, USA, 02139
[5]Trauma Surgery, Cleveland MetroHealth Hospital, Cleveland, OH, USA, 44106

ABSTRACT

Biodegradable polymers with high mechanical strength, flexibility and optical transparency, optimal degradation properties and biocompatibility are critical to the success of tissue engineered devices and drug delivery systems. In this work, microfluidic devices have been fabricated from elastomeric scaffolds with tunable degradation properties for applications in tissue engineering and regenerative medicine. Most biodegradable polymers suffer from short half life resulting from rapid and poorly controlled degradation upon implantation, exceedingly high stiffness, and limited compatibility with chemical functionalization. Here we report the first microfluidic devices constructed from a recently developed class of biodegradable elastomeric poly(ester amide)s, poly(1,3-diamino-2-hydroxypropane-co-polyol sebacate)s (APS), showing a much longer and highly tunable in vivo degradation half-life comparing to many other commonly used biodegradable polymers. The device is molded in a similar approach to that reported previously for conventional biodegradable polymers, and the bonded microfluidic channels are shown to be capable of supporting physiologic levels of flow and pressure. The device has been tested for degradation rate and gas permeation properties in order to predict performance in the implantation environment. This device is high resolution and fully biodegradable; the fabrication process is fast, inexpensive, reproducible, and scalable, making it the approach ideal for both rapid prototyping and manufacturing of tissue engineering scaffolds and vasculature and tissue and organ replacements.

INTRODUCTION

One of the principal challenges in tissue engineering is the requirement for a vasculature to support oxygen and nutrient transport within the growing tissue. One avenue for achieving this goal is the formation of a microfluidic network within the tissue engineering scaffold; an initial proof of principle for this concept was demonstrated using nondegradable PDMS as the substrate for endothelialized microfluidic networks.[1] Early demonstrations of biodegradable microfluidic devices capable of supporting microvascular networks were reported by Armani and Liu,[2] King et al.,[3] and Liu and Bhatia.[4] The first of these reports required the insertion of a nondegradable metallic layer for bonding the degradable PLGA films; the latter two were constructed solely from biodegradable PLGA. These PLGA-based structures suffered from excessive mechanical stiffness and sudden changes in mass and mechanical strength during the resorption process, spurring development of alternative biodegradable materials with more optimal mechanical and resorption properties. The first of these was a biodegradable elastomer poly(glycerol sebacate) (PGS).[5] The relatively short half-life of PGS led to further development of APS as a tunable biodegradable elastomer for tissue engineering applications; here we present the first report of a microfluidic network capable of serving as an intrinsic vasculature formed from the APS polymer.

To develop a longer-lasting scaffold substrate than PGS while preserving its excellent chemical and mechanical properties, a new class of biodegradable elastomers, poly(ester amide), poly(1,3-diamino-2-hydroxypropane-co-polyol sebacate) (APS) with tunable degradation rates was developed.[6] This class of APS materials encompasses a wide range of chemical compositions. These compositions have been shown to be highly biocompatible, with primary hepatocyte culture exhibiting cell functionality over extended periods without the need for deposition of protein coatings on the surface. In addition, the APS surface is amenable to nanostructuring to provide topographic features representative of the cell microenvironment.[7] In this work, we present microfabrication results for microfluidic channel networks as well as in vitro degradation data as a function of selective enzyme triggers for two APS polymer compositions. These data demonstrate that the APS system represents a novel class of polymers for the development of microfluidic scaffolds for tissue engineering applications.

EXPERIMENTAL METHODS

Synthesis of poly(1,3-diamino-2-hydroxypropane-co-glycerol sebacate) elsatomers

All materials were purchased from Sigma Aldrich (St. Louis, MO, USA) and used as received unless otherwise specified. A round bottom flask was charged with 0.06 mol of 1,3-diamino-2-hydroxy-propane (DAHP), 0.03 mol glycerol (G), and 0.09 mol of sebacic acid (SA) to produce a molar ratio of 2:1:3 of DAP:G:SA, respectively to produce 2-1 APS, while 1-2 APS was synthesized with a molar ratio of 1:2:3. The reactants were heated under an argon blanket at 130 °C for 3 h. The pressure was then dropped to approximately 50 mTorr and the contents were allowed to react for 10 h at 130 °C. The product was then stored under a desiccant environment until further use.

Microfabrication of microfluidic scaffold using reverse silicon molding

The silicon mold was designed and created as previously described by King *et al.*[3] Briefly, prior to replica molding of APS, a sacrificial maltose release layer was spin-coated on the silicon master. Photolithographically patterned silicon masters were cleaned using piranha solution (Mallinckrodt, St. Louis, MO) and oxygen plasma-cleaned (March, St. Petersburg, FL) at 250 mTorr and 200 W for 45 seconds. A 60% (w/w) solution of maltose (Sigma, St. Louis, MO) in water was spin-coated at 2500 revolutions per minute for 30 s. The maltose layer was pre-baked on a hot plate at 95 °C for 120 seconds. 5.00 ± 0.05 g of APS prepolymer was melted at 170 °C and applied to the wafers for replica molding and smooth sheet formation. The prepolymer was cured at 170 °C for 48 h under 50 mTorr of vacuum, which produced a crosslinked sheet in which a portion of the hydroxyl and carboxylic acid functional groups remained. The APS sheets were delaminated by statically incubating the polymer-master system in doubly distilled water (ddH2O) at 60 °C for 24 h beginning immediately after polymer curing. Diffusion of water between the polymer/silicon interfaces led to maltose dissolution and eventual delamination. Sheets were trimmed and punched to achieve appropriate fluidic connections between layers. Microfluidic APS layers were stacked, aligned, and bonded together simultaneously by oxygen plasma treatment for surface activation, followed by curing the polymer at 170 °C for 24 h under 50 mTorr of vacuum. An additional layer of PDMS was laminated onto the device for tubing stabilization. Once the final curing step was completed, silicone tubing (1/16 in. inner diameter, 1/8 in. outer diameter, Cole-Parmer) was inserted into the devices in a sterile environment. Luer-Lok connections were inserted into the tubing, and the base of the connections was sealed with epoxy (McMaster-Carr).

Microfabrication of microfluidic scaffold using reverse PDMS molding

The SU-8 patterned wafer was designed and created as previously described by Bettinger et al.[8] Prior to replica molding of PDMS, fluoropolymer was coated as passivation layer on the master. 25.00 ± 0.05 g of PDMS prepolymer was applied to the wafers for PDMS replica molding. The PDMS was crosslinked at 70°C for 4 h. Prior to replica molding of APS on PDMS, similar maltose coating procedures to the previous section were employed. 5.00 ± 0.05 g of APS prepolymer was melted at 170 °C and applied to the wafers for replica molding and smooth sheet formation. The prepolymer was cured at 170 °C for 48 h under 50 mTorr of vacuum, which produced a crosslinked sheet in which a portion of the hydroxyl and carboxylic acid functional groups remained. The APS sheets were delaminated by statically incubating the polymer-master system in doubly distilled water (ddH2O) at 60 °C for 24 h beginning immediately after polymer curing. Diffusion of water between the polymer/PDMS interfaces led to maltose dissolution and eventual delamination. Devices were assembled using similar techniques as described in the previous section.

Degradation of 2-1 APS and 1-2 APS

Cylindrical polymer slabs (n=3) with dimensions of 0.5mm × 6mm (T × D) and weighing ~20mg each were incubated at 37°C in the following degradation media: (1) Dulbecco's Phosphate Buffer Saline (DPBS); (2) 10 U mL^{-1} of protease I from bovine pancreas in DPBS; (3) 10 U mL^{-1} lipase II from porcine pancreas in DPBS. Buffer and enzyme solutions were exchanged every 2 days and dry mass loss measurements were made at specific time points. Samples were washed in ddH$_2$O, incubated in ethanol and ddH$_2$O for at least 24 h. Dry mass were weighed for mass loss comparison.

DISCUSSION

Microfabrication of microfluidic scaffold

The basic technical approach for designing and constructing a microvascularized scaffold device here is similar to that reported for PLGA,[3] PGS[8-9] and silk fibroin.[10-11] Briefly, a simple microchannel network design was generated based on principles of microvascular flow, pressure and wall shear stress.[12] This microchannel network is based upon design principles including uniform flow and distribution of oxygen and nutrients throughout a scaffold layer, as well as physiologic levels of pressure drop and wall shear stress within the various vessel sizes in the network. The network was translated into a photolithographic layout for fabrication, with the intent of producing a single-layer microvascular network.

For PGS, the optimal surface coating was found to be sucrose dissolved in water, but process development with APS determined that the temperatures involved in curing resulted in unacceptably high levels of sucrose caramelization during the process. Therefore, we investigated alternative sacrificial sugar coatings and determined that maltose has a caramelization temperature about 20 $^{\circ}$C higher than sucrose. Therefore, maltose coatings were deposited onto the silicon masters to assist in delamination of APS films, and APS was cast onto the silicon master molds and then lifted off as a free-standing film. A channel layer was bonded to a flat APS layer and the layers joined at elevated temperature and pressure. The challenge for layer bonding is to form a strong, irreversible and leakproof bond between films, without raising the temperature or pressure so high as to deform or collapse the microchannels or other high resolution structures. In order to prepare the bonded devices for future flow testing and cell seeding, silicone tubing was attached to the APS microfluidic scaffold to demonstrate flow through the network at physiological pressures (0-180 mmHg). As seen in Figure 1, robust microfluidic devices were built using APS, and exhibited good flow behavior. It was shown that unlike softer alternative materials, the strength of APS was sufficient to support maintenance of the channel height, as well as suitable flow dynamics.

Figure 1. a) Full microfluidic channel device in APS flowing with red dye. b) Microfluidic channels mimicking blood vessel bifurcated networks built in APS flowing with red dye.

Degradation of 2-1 APS and 1-2 APS

The comparison of degradation properties between 2-1 APS and 1-2 APS showed that by varying the ratio of diamine to glycerol (2:1 for 2-1 APS and 1:2 for 1-2 APS) the degradation rate of APS is tunable as shown in Figure 2. The higher the diamine content, the slower the polymer degrades, and the less the polymer responds to lipase solution as a catalyst for degradation.

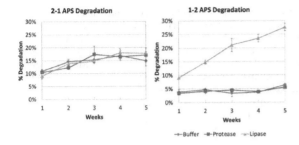

Figure 2. a) 2-1 APS degradation over 5 weeks in buffer, protease, and lipase solution. B) 1-2 APS degradation over 5 weeks in buffer, protease, and lipase solution.

It is also shown in Figure 3 that the polymer undergoes surface erosion as oppose to bulk resorption, which is beneficial in maintaining functionality of the device during degradation.

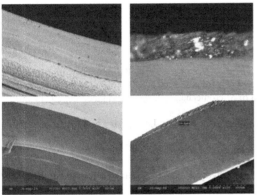

Figure 3. a) Microscopic image of APS in buffer solutions for 4 days. b) Microscopic image of APS in lipase solution for 4 days. c) SEM of APS in buffer solution for 4 days. d) SEM of APS in lipase solution for 4 days.

CONCLUSIONS

The concept of biodegradable microfluidic devices was first introduced using PLGA in by Armani and Liu[2] and King et al.,[3] but these device constructs exhibit a high degradation rate, low elasticity, and concerns regarding immune and inflammatory response in bulk format. More recently, Fidkowski et al.[9] and Bettinger et al.[8] constructed PGS-based devices that exhibit much more desirable mechanical properties for implantation, but the relatively high degradation rate of the polymer limits its applicability for certain long-lasting applications as an implant. In the present study, we report on the development of a specific process for biodegradable microfluidic devices constructed from APS, an alcohol-based poly(ester amide) elastomer, which possesses a much lower degradation rate but retains elastomeric properties for tissue scaffold applications. The process reported here is capable of forming microfluidic channels on a wafer-scale with strong layer-to-layer bonds and well-preserved microscale architecture. Robust flow conditions within the devices were developed. The polymer also showed selective degradation toward lipase solution. It was also observed that by varying the amine to glycerol ratio, the degradation properties of APS can be tuned. The biocompatibility of APS *in vivo* by Bettinger et al.[13] demonstrated that the material could be used for resorbable tissue engineering devices for the purpose of drug delivery and regenerative medicine. The next steps for this investigation will focus on the seeding of endothelial cells to form a functional microvascular network, and implantation studies to explore in vivo biocompatibility, host integration and long-term degradation properties for this scaffolding material.

ACKNOWLEDGMENTS

Funding for this work, provided by Draper Laboratory and by the American Association for the Surgery of Trauma, is gratefully acknowledged. We are indebted to G.C. Engelmayr, C.J. Bettinger, L.E. Freed, J. Hsiao, and E. Kim for many useful discussions and assistance with the instrumentation and polymer synthesis.

REFERENCES

1. J. T. Borenstein, H. Terai, K. R. King, E. J. Weinberg, M. R. Kaazempur-Mofrad and J. P. Vacanti, Biomedical Microdevices **4** (3), 167-175 (2002).
2. D. K. Armani and C. Liu, Journal of Micromechanics and Microengineering **10** (1), 80-84 (2000).
3. K. R. King, C. C. J. Wang, M. R. Kaazempur-Mofrad, J. P. Vacanti and J. T. Borenstein, Advanced Materials **16** (22), 2007-12 (2004).
4. V. Liu Tsang, A. A. Chen, L. M. Cho, K. D. Jadin, R. L. Sah, S. DeLong, J. L. West and S. N. Bhatia, FASEB J **21** (3), 790-801 (2007).
5. Y. D. Wang, G. A. Ameer, B. J. Sheppard and R. Langer, Nature Biotechnology **20** (6), 602-606 (2002).
6. C. J. Bettinger, J. P. Bruggeman, J. T. Borenstein and R. S. Langer, Biomaterials **29** (15), 2315-2325 (2008).
7. C. J. Bettinger, Pure and Applied Chemistry **81** (12), 2183-2201 (2009).
8. C. J. Bettinger, E. J. Weinberg, K. M. Kulig, J. P. Vacanti, Y. D. Wang, J. T. Borenstein and R. Langer, Advanced Materials **18** (2), 165-9 (2006).
9. C. Fidkowski, M. R. Kaazempur-Mofrad, J. Borenstein, J. P. Vacanti, R. Langer and Y. D. Wang, Tissue Engineering **11** (1-2), 302-309 (2005).
10. J. T. Borenstein, M. M. Tupper, P. J. Mack, E. J. Weinberg, A. S. Khalil, J. Hsiao and G. Garcia-Cardena, Biomedical Microdevices **12** (1), 71-79 (2010).
11. C. J. Bettinger, K. M. Cyr, A. Matsumoto, R. Langer, J. T. Borenstein and D. L. Kaplan, Advanced Materials **19** (19), 2847-50 (2007).
12. J. T. Borenstein, E. J. Weinberg, J. P. Vacanti and M. R. Kaazempur-Mofrad, in *Micro and Nanoengineering of the Cell Microenvironment*, edited by A. Khademhosseini, J. T. Borenstein, M. Toner and S. Takayama (Artech House, Boston, 2008).
13. C. J. Bettinger, K. M. Kulig, J. P. Vacanti, R. Langer and J. T. Borenstein, Tissue Engineering Part A **15** (6), 1321-1329 (2009).

Mater. Res. Soc. Symp. Proc. Vol. 1299 © 2011 Materials Research Society
DOI: 10.1557/opl.2011.63

Measurements of Resonance Frequency of Parylene Microspring Arrays Using Atomic Force Microscopy

C. Gaire[1], M. He[1], A. Zandiatashbar[2], P.-I. Wang[1], R. C. Picu[2], G.-C. Wang[1] and T.-M. Lu[1]
[1]Department of Physics, Applied Physics and Astronomy, Rensselaer Polytechnic Institute, 110 8th Street, Troy, NY, 12180, U. S. A.
[2]Department of Mechanical, Aeronautical and Nuclear Engineering, Rensselaer Polytechnic Institute, 110 8th Street, Troy, NY, 12180, U. S. A.

ABSTRACT

A mechanical vibration system was made by sandwiching an array of parylene-C microsprings between two flat plates of Si. This system was driven mechanically in forced oscillation using a piezo transducer attached to the bottom Si plate. An atomic force microscope was used to record the displacement of the top plate in both the contact and non-contact modes. At the resonance, the system was observed to give large vertical displacement amplitude of up to 100 nm with a Q-factor of up to 900.

INTRODUCTION

Recent advents of micro- or nanospring arrays for optical interferometry [1], pressure sensing [2] and electromechanical actuation [3,4] have generated a significant interest in the use of such structures as elements of micro- and nanoelectromechanical systems (MEMS and NEMS). Most of the current MEMS or NEMS elements are made of Si microstructures that inherit the advantages of fabrication by planar processing techniques used in silicon microelectronic technology [5-7]. The use of microstructures made of polymers such as parylene as elements of MEMS or NEMS is less developed to date [8,9]. Due to its low elastic modulus, high structural flexibility, chemical robustness and ease of fabrication, parylene microstructures could play an important role in future MEMS or NEMS devices. The ability of an accurate measurement of natural frequency of such structures is required to gain control over the desired precision in the mechanical motion. Hence, resonance frequency measurement is one of the most important prerequisite of MEMS or NEMS elements.

In this article, we present the results of resonance frequency measurement of a mechanical vibration system composed of an array of parylene-C microsprings using an atomic force microscope (AFM). The system contains several millions microsprings sandwiched between two Si plates. A piezoelectric transducer is used to drive the system from the bottom plate and the AFM is used to record the displacement of the top plate.

EXPERIMENT

Parylene spring growth

The growth of parylene-C microsprings was carried out by oblique angle deposition [10]. We have employed a substrate swing rotation scheme during deposition to control the size and uniformity of the springs [11,12]. In short, the parylene vapor sublimated at 190 °C was passed through a pyrolysis furnace at 680 °C to convert to parylene-C monomers. This monomer flux was guided by a nozzle [10] and was deposited on Si(001) wafer at an angle of ~ 85° with respect to the substrate normal. The parylene-C monomers polymerize upon deposition onto the substrate [13]. Due to the shadowing effect and limited mobility of deposited material, isolated islands are formed on the surface and serve as seeds for the growth of microsprings. The substrate was rotated in swing motion at 10 rpm within the azimuthal angle of 90° without changing the deposition angle with respect to the substrate normal. Each turn of the spring consists of four discrete units grown by successive deposition of arms with substrate turned by 90° after the completion of each arm. These springs had the following dimensions: wire diameter, $t = (2.3\pm0.6)$ μm; spring pitch, $h = (7.5\pm0.3)$ μm, coil diameter $D = (5.0\pm0.5)$ μm, rise angle of ~ 30°, and number of turns, $n = 4$.

Mechanical vibration system

The mechanical vibration system was made by bonding the sample to a Si piece (dimensions: 12.5 mm × 12.7 mm × 725 μm; mass: 268 mg) through ~2 μm thick epoxy siloxane polymer (obtained from Polyset Co. Inc.) [14] spin-coated on one side and pressed from the other side with a force of 2 N (pressure =12.6 kN/m^2). The bonded sample is annealed in inert gas environment at 160°C for one hour to remove the residual solvent inside the epoxy siloxane polymer film and harden this adhesive layer. The lower plate is glued to a piezoceramic sheet and the upper plate is free to oscillate in unison to the microsprings grown on the bottom plate. The piezoceramic sheet was 191 μm thick lead zirconate titanate (PZT) with nickel electrodes (Piezo Systems Inc., Cambridge, MA). The normal strain coefficient of the sheet was 390 pm/V. The PZT sheet was sinusoidally driven with peak-to-peak voltage of 1.0–7.0 V (peak-to-peak amplitude: 0.39-2.73 nm) in two frequency ranges: 0.001–0.1 MHz (using SR830 Lock-in amplifier, Stanford Research Systems, Sunnyvale, CA) and 0.1–100 MHz (using HP 8656B function generator).

Spring constant determination

The stiffness of an individual spring was determined using nanoindentation (NanoTest, MML, Wrexham, UK) on parylene sample without the top Si plate. A flat punch nanoindentation tip of diameter 50 μm was used to indent on 24 different locations. Based on the pitch of the spring array, diameter of the spring wire and the size of the indenter tip, it was estimated that this tip indented 80±10 springs in each indentation. The individual spring constant is determined to be $k =21\pm5$ N/m by dividing the total stiffness by the number of springs loaded in each location. The loading-unloading curves were not perfectly overlapping to each other, so the stiffness was determined from the unloading portion of the nanoindentation curves. This spring constant will be used to estimate the theoretical resonance frequency of the vibrating system.

Resonance frequency measurement

An XE-100 Series AFM (PSIA Inc., Santa Clara, CA) was used to measure the vertical amplitude of the top plate while the actuation frequency is swept from 1 kHz to 100 MHz. The displacement of the top plate was measured in both non-contact and contact modes of the AFM. In non-contact mode, the displacement of the top plate was measured dynamically by vibrating the cantilever at a frequency slightly larger than its natural frequency at a distance from the sample proportional to the set point value [15]. In our case, a set point value of 6 nm was used with a cantilever with natural frequency = 165 kHz and spring constant = 42 N/m. In the contact mode, the upper plate was pushed by the cantilever with a force of 1 nN and the displacement of the plate was monitored by recording the cantilever deflection as a function of the driving frequency. The cantilever used in the contact mode had a spring constant of 0.01 N/m and natural frequency of 12 kHz. Both the non-contact and contact modes of AFM were used to measure mechanical resonance of microresonators [16], single nanorods of Si [15], coiled carbon nanotubes [17] and NEMS oscillators [18]. The contact mode measurement is generally used to isolate any inherent dynamic coupling of AFM microcantilevers in non-contact mode with the motion of the sample [19].

RESULTS AND DISCUSSION

A cross section scanning electron microscope (SEM) image of 4-turn parylene-C springs is shown in Fig. 1(a). The total number of springs in the given sample was found to be $N = 3.33 \times 10^6$, and was estimated by sampling the number of springs seen in a given area of top view SEM images (not shown here). A schematic of the resonance frequency measurement set up is shown in Fig. 1(b). The piezo was driven using a function generator rendering the microspring array to a forced oscillation.

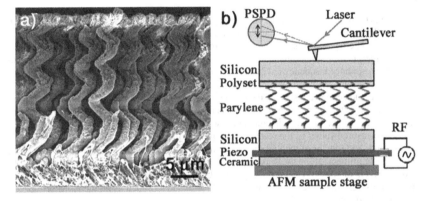

Figure 1: (a) Cross section SEM image of the parylene-C microspirals (b) Schematic of the mechanical vibration system and resonance frequency measurement method. PSPD and RF stand for position sensitive photo diode (in AFM) and radio frequency generator, respectively.

Figure 2 shows the amplitude vs. driving frequency of the system measured in contact and non-contact mode of the AFM with a driving signal of 0.9 nm peak-to-peak amplitude. When the driving frequency is equal to the natural frequency of the spring-mass system, large vertical amplitude was observed. With driving signal amplitude of 0.9 nm, the amplitude of vibration of the system in resonance was found to be approximately 30 nm when measured in contact and non-contact modes of AFM. Vibration amplitude as large as 100 nm was observed for the largest driving signal of 7 V (or 2.73 nm peak-to-peak) used.

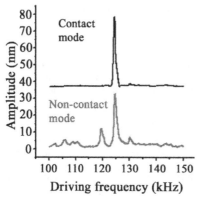

Figure 2: Amplitude as a function of driving frequency of the vibration system measured in both contact and non-contact mode of AFM. The baseline for the contact mode curve is offset by 35 nm for a better visibility.

In contact mode, the measured resonance frequency was 124.5 kHz with a quality factor (Q-factor) of 896. In non-contact mode, the measured resonance frequency was 124.8 kHz with a Q-factor of 588. There was a slight difference (approximately 300 Hz) in the natural frequency of the sample measured in non-contact and contact modes. The difference is about 0.2% of the peak value of the resonance frequency (~124 kHz). The non-contact mode measurement also yielded satellite peaks around the main resonance peak. These are most likely due to extrinsic causes: either from the noise in the system or the dynamic nature of the non-contact mode measurement method. It is clear that the resonance peak is cleaner (larger Q-value and free of satellite peaks) when the measurement was done in contact mode as opposed to the non-contact mode of the AFM. The measured values of the resonance frequency, the full width at half maximum (FWHM) and the Q-factor of the vibration system are summarized in Table 1. In order to examine the legitimacy of the measurements, we performed similar experiments on microcantilevers of known resonance frequency. In both the contact and the non-contact mode measurements, resonance frequencies within 0.1% to that determined using the AFM set-up (optical method) were observed.

Table 1: Summary of the measured parameters (resonance frequency, FWHM and Q-factor) in non-contact and contact modes of AFM

Measurement method	Resonance frequency (Hz)	FWHM (Hz)	Q-factor
Contact mode	124,470	873	896
Non-contact mode	124,800	1,333	588

The theoretical resonance frequency of the vibrating system was estimated using

$$f = \frac{1}{2\pi} \sqrt{\frac{Nk}{m^*}}$$
(1)

where N is the number of microsprings acting in parallel, k is the stiffness of an individual spring and m^* is the effective mass of the system (spring array and top plate). Based on the dimensions of the top Si plate, the microsprings and the bonding layer of polyset, the effective mass of the vibration system was approximated by the mass of the silicon plate ($m^* = 268$ mg). The total number of microsprings acting in parallel for this particular system, N was estimated to be 3.33×10^6. From the nanoindentation experiments, the elastic spring constant, k, of the individual spring was determined to be $k = 21 \pm 5$ N/m. So the estimated resonance frequency of the system turns out to be (81.3 ± 9.7) kHz. This value of the resonance frequency is approximately 35% less than the measured value of ~124 kHz. The large error in the estimated value of the resonance frequency could originate from errors in the estimated spring constant (25%) because of the spring-spring interaction during the loading-unloading cycles, the estimated number of springs sandwiched between the silicon plates, the effective mass of the system as well as from the estimated number of springs (12%) during the nanoindentation experiment. Also, the fact that these springs were grown on a large sample, they may not be uniform over the entire domain. This could also add to the uncertainty. By growing samples in patterned substrates, the error in the estimation of number of springs (sandwiched and/or nanoindented) and hence the error in the estimated frequency can be lowered.

CONCLUSION

A mechanical vibration system was made using an array of parylene-C microsprings and its resonance frequency was measured using AFM. At the resonance, the system exhibited displacement amplitudes on the order of 10-100 nm with a Q-factor of several hundreds. No significant difference in resonance frequency is observed between the contact and non-contact AFM measurements, indicating the coupling between the AFM cantilever and the sample is negligible. A large Q-factor indicates that the system could find applications in MEMS devices.

ACKNOWLEDGMENTS

This work was partially supported by the National Science Foundation of USA (grant number 0506738) and Semiconductor Research Corporation.

REFERENCES

[1]   G. D. Dice, M. J. Brett, D. Wang and J. M. Buriak, *Appl. Phys. Lett.* **90**, 253101 (2007).
[2]   S. V. Kesapragada, P. Victor, O. Nalamasu and D. Gall, *Nano Lett.* **6**, 854 (2006).
[3]   J. P. Singh, D.-L. Liu, D.-X. Ye, R. C. Picu, T.-M. Lu and G.-C. Wang, *Appl. Phys. Lett.* **84**, 3657 (2004).
[4]   G. Zhang and Y. Zhao, *J. Appl. Phys.* **95**, 267 (2004).
[5]   V. Y. Prinz, V. A. Seleznev, A. V. Prinz and A. V. Kopylov, *Sci. Technol. Adv. Mater.* **10**, 034502 (2009).

[6]   E. G. Kostsov, *Optoelectronics, Instrumentation and Data Processing* **45**, 189 (2009).

[7]   L. Dong, L. Zhang, D. J. Bell, D. Grützmacher and B. J. Nelson, *J. Phys.: Conf. Series* **61**, 257 (2007).

[8]   D. W. Panhorst, V. LeFevre and L. K. Rider, *Ferroelectrics* **342**, 205 (2006).

[9]   S. Sung, J. G. Lee, B. Lee and T. Kamg, *J. Micromech. Microeng.* **13**, 246 (2003).

[10]   G. Demirel, N. Malvadkar and M. C. Demirel, *Thin Solid Films* **518**, 4252 (2010).

[11]   D.-X. Ye, Z.-P. Yang, A. S. P. Chang, J. Bur, S. Y. Lin, T.-M. Lu, R. Z. Wang and S. John, *J. Phys. D: Appl. Phys.* **40**, 2624 (2007).

[12]   M. M. Hawkeye and M. J. Brett, *J. Vac. Sci. Technol. A* **25**, 1317 (2007).

[13]   J. B. Fortin and T.-M. Lu, *J. Vac. Sci. Technol. A* **18**, 2459 (2000).

[14]   P.-I. Wang, O. Nalamasu, R. Ghoshal, R. Ghoshal, C. D. Schaper, A. Li and T.-M. Lu, *J. Vac. Sci. Technol. B* **26**, 244 (2008).

[15]   T. C. Parker, F. Tang, G.-C. Wang and T.-M. Lu, *Sensors and Actuators A: Physical* **148**, 306 (2008).

[16]   S. Ryder, K. B. Lee, X. Meng and L. Lin, *Sensors and Actuators A* **114**, 135 (2004).

[17]   A. Volodin, D. Buntinx, M. Ahlskog, A. Fonseca, J. B. Nagy and C. V. Haesendonck, *Nano Lett.* **4**, 1775 (2004).

[18]   B. Ilic, S. Krylov, L. M. Bellan and H. G. Craighead, *J. Appl. Phys.* **101**, 044308 (2007).

[19]   F. Gittes and C. F. Schmidt, *Eur. Biophys. J.* **27**, 75 (1998).

Mater. Res. Soc. Symp. Proc. Vol. 1299 © 2011 Materials Research Society
DOI: 10.1557/opl.2011.376

Gold in Flux-less Bonding: Noble or not Noble.

Marco Balucani[1], Paolo Nenzi[1], Fabrizio Palma[1], Hanna Bandarenka[2], Leonid Dolgyi[2], Aliksandr Shapel[2]

[1]Department of Information Engineering, Electronics and Telecommunications, Sapienza – Università di Roma, Via Eudossiana, 18 – 00184 Roma (Italy)
[2]Micro- and nano-electronics Department, Belarussian State University of Informatics and Radioelectronics, P. Brovka, 6 – 220027 Minsk, Belarus

ABSTRACT

This work highlights the solder joints reliability issues emerged during the development of a novel compliant contacting technology. The peculiar process in this technology is a mechanical lifting procedure in which a pulling force is exerted onto 63Sn-37Pb (eutectic) solder joints (realized by a flux-less thermo compression process), releasing metal traces from the substrate, to form free standing vertical structures. Since joints mechanical characteristics are critical for the successful fabrication of contacts, different bonding conditions (inert or forming atmosphere, temperature rates) and surface finishing (electroplated gold and preformed solder) have been tested. SEM and EDX analyses have been performed on failing joints to investigate failure causes and classify defect typologies. A constantly higher failure rate (percent number of failing joints) has been observed on gold finished surfaces. Analyses proved that such unusual rate was due to contamination of gold surface left by additives in the plating bath and to the embrittlement caused by gold diffusion into molten solder. Plating additives contamination reduces the wettability of gold surfaces. Concentration values of 3 wt.% for gold, considered safe for surface mount applications, caused embrittlement in solder bumps of 20-40 μm diameters.

INTRODUCTION

The continuous scaling of integrated circuit features affects the dimensions of bonding pads and their pitch. This in turn requires that packaging and probing technologies scales as well. Recently a construction of a novel compliant contacting technology [1] applicable to both integrated circuits testing (wafer level probe card) and packaging was proposed [2]. In the assembly steps of the technology, the silicon substrate holding the metal wires forming the contacting structures is bonded to a ceramic (alumina) substrate holding the signal redistribution layer by means of 63Sn-37Pb solder in a thermo-compression process. Once bonded, the metal wires are released from the substrate "pulling" the two substrates away and keeping them at a prescribed distance. The space between the two substrates (enclosing free-standing wires) is filled with a thermoset polymer by capillarity. As the polymer is cured, the silicon substrate is removed by XeF$_2$ etching in order to expose the metal wire ends (the contacting tips).

Solder joint design

In this technology the solder joint is subject to a pulling force exerted to release the metal wire from the substrate where it has been plated. The adhesion between substrate and metal has to be carefully controlled to ensure separation of the metal from the substrate without breaking the joint or the metal itself. Adhesion control is achieved by plating wires over a porous silicon adhesion layer [3] obtained from silicon by anodization process. Porosities of 30% to 50% (depending on the metal thickness) of the porous silicon layer are required to achieve a 1 MPa adhesion. The value of 1 MPa has been determined experimentally to be high enough for reliable processing of wafers. As the chosen value of adhesion is 20 times lower than the yield strength of solder (i.e. 27.2 MPa [4]), it is expected that, by pulling, the porous layer will break, releasing the metal pad from silicon and peeling off the wire, without breaking the joint. The technology has been tested on bonding pads having 30 μm diameter and 70 μm pitch. On such small pads, electroplating is the only available technology for solder application. Pads on silicon substrates have been plated with 20-40 μm thick layer of solder. Pads on ceramics were finished with 0.25 μm electroplated Au (over 2 μm Ni) or with 20 μm 63Sn-37Pb (over 0.5 μm Ni). The volume of the deposited solder is the key parameter as it affects the mechanical and dimensional characteristics of the solder joint and, ultimately, the reliability of the assembly. This is particularly true when gold is used as surface finish [5-9] because a gold concentration of 5 wt.% or lower causes embrittlement [10].

Bonding processes

Thermo-compression bonding has been used for this technology. Compressive force, applied to the assembly during the reflow process, is needed to break the thin oxide layer that covers the solder, so that fresh solder can come into contact with the pad to wet it, and to compensate the height differences arising from electroplating process variability. In this technology, fluxing agents are not applicable because the residues cannot be cleaned without the risk of releasing the structures. Bonding processes were executed on the two different surface finishing in three ambient conditions: air, nitrogen and reducing atmosphere. Different test runs were executed on manual and automated die bonders (with formic acid reducing atmosphere) to find the optimal force and reflow profiles. The force applied during the reflow process has found to be a critical parameter of this technology due to the fine pitch (10 μm gap between pads). The force must be kept small to not spread-out the liquid solder from pad area (bridging adjacent pads) and, at the same time, must be sufficient to break the oxide layer over the bumps and wet the substrate pad. The use of formic acid has been studied [11] as improvement factor in the self-alignment chip bonding and has proven to be necessary in this technology to achieve the total wetting of substrate pads at lower forces, necessary to reduce the risk of solder bridges.

The optimal process parameters are: force 0.5 g/bump, preheat temperature 150°C, peak temperature 240°C, dwell time 5 s, heating ramp 2°C/s, cooling ramp 2°C/s, formic acid 10 l/min at 200°C. The total time above liquidus for this process is 75 s, the minimum that guarantees full pad wetting. Minimizing the time above liquidus is necessary to keep gold dissolution into solder at minimum. With these parameters, the highest achieved yield, were 89% for preformed solder finishing and 56% electroplated gold finishing; the latter has been extensively studied to understand the causes of low wettability and failures.

RESULTS AND DISCUSSION

The gold wettability issue was attributed to the plating additives; Auger spectroscopy revealed the presence of thallium oxide (thallium is an additive used in Au plating baths).

This thin oxide layer (extending for 5 nm) has been removed by ion milling prior to bonding and this resulted in an increase in yield, but to a value still lower then the one of preformed solder finishes. This difference has been the subject of an in-depth study to discover the causes and find a possible solution, as gold finish was preferred over preformed solder.

Figure 1 shows part of a sample, where is possible to see all the revealed defect mechanisms that affected the bonding and pulling processes. Arrows in the center shows matching pads, and arrows in the magnifications mark the areas (1×1 μm^2) where EDX analysis has been done. Squares (solid and dotted) correspond to 3×3 μm^2 and 10×10 μm^2 areas where the average quantitative analysis of the Au, Sn and Pb has been computed. The analyses results are presented in fig. 3 and fig. 4.

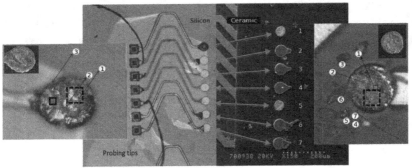

Figure 1. Detail of a gold finished sample (silicon left, ceramic right) after pulling process. Pad 1 (both sides) is magnified to show micro and quantitative analyses areas. The visible film on optical photo (left) is a residue of flux used for solder reflow. The probing tips (at the left) are still attached to the surface even if corresponding wires have been completely released.

All examined defects fall in 3 categories: solder joint failure without release of wire from the silicon substrate (pad 1); solder joint failure with release of the metal wire (pad 5); silicon cratering with or without wire break (pads 2,6,7). The missing metal wires (on pads 3 and 4) are not classified as failures as the wires broke during handling of the samples.

Pad number 1 is representative of the most frequent failure encountered during testing and has been object of SEM (Scanning Electron Microscope) and EDX (Energy-Dispersive X-ray spectroscopy) analysis to reveal the microstructure and composition of the failing joint. The SEM photos in fig. 2 show the fracture on the ceramic side, (a) and (b) and silicon side, (c) and (d). The fracture is located between 3 and 6 μm from the pad surface and the cause is attributed to solder contamination by intermetallic compound that embrittle the joint. The morphology of AuSn IMC (intermetallic compounds) is evident [12] in both sides of the fracture (fig. 3(b) and (d)). While EDX data cannot be readily used to quantify the concentration of material in the specimen, it is evident that a ductile fracture happens where Au is absent (e.g. position 1 on silicon side and position 3 on ceramic side) and brittle fracture happens at high Au concentrations (positions 2 and 3 on the rim of the pad). EDX data in fig. 3 (ceramic side), reveals the presence of Au on the entire surface, indicating that gold has diffused into solder, concentration in the dotted square is >35 wt.%. Position 3, where the fracture appears ductile, corresponds to a Pb-rich area. EDX on silicon side of pad 1 (fig. 4) reveals that Au concentration is >28 wt.% inside the brittle fracture (3×3 μm^2 area) where the AuSn intermetallics can be recognized in the morphological analysis (fig. 2(d)) and <6 wt.% in the ductile Pb-rich fracture (10x10 μm² area). Analyses run on other pads show similar results.

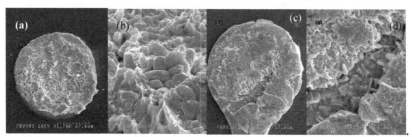

Figure 2. SEM images of pad 1; ceramic side (a), magnification of pad1 to reveal intermetallic compounds (b); silicon side (c) and magnification of fracture on silicon side (d).

The cause of low yield is the gold poisoning of the solder joint. Gold finish, used to protect lower metal layers from oxidation and contamination, is known to influence the soldering process and joint reliability when 63Sn-37Pb solder is used.

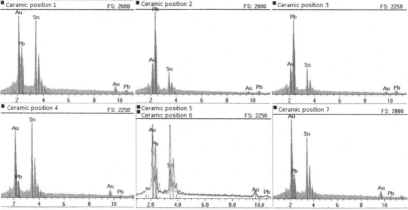

Figure 3. EDX analysis on pad 1(ceramic). Position numbers reflect the one in fig. 1. The normalized results are: Pb 29 wt.%, Au 36 wt.% and Sn 35 wt.%.

The concentration of Au affects tin lead solder in the following [9]: fluidity, wettability and spread, mechanical properties (solder ductility is drops rapidly as the Au concentration exceeds 7 wt.%), microstructure (at concentrations greater than 1 wt.% solder becomes detectable in the microstructure as needle-shape phase). There is no general agreement on the maximum Au concentration that can be tolerated in a solder joint, with the most restrictive constraint set to 3 wt.% and most of the authors suggest to test Au influence in every specific case.

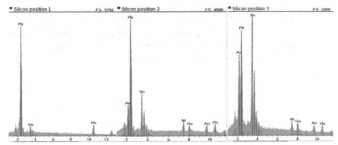

Figure 4. EDX analysis on pad 1 (silicon). Position numbers reflect the one in fig. 1. The normalized results are: Pb 76 wt%, Au 6 wt.%, Sn 18 wt.% in the 10x10 μm² area and, Pb 26 wt.%, Au 28 wt.%, Sn 46 wt.% in the 3x3 μm² area.

When the content of Au is excessive, the following phenomena may happen: solder joint fracture due to embrittlement, void creation (Kirkendall voids), and microstructure coarsening. In this specific case, the 3 wt.% threshold is not applicable and the cause of this additional detrimental effect lies in the interface kinetics of Au with solder where both dissolution of Au into solder and interfacial reactions, contribute to the formation of brittle AuSn intermetallic compounds. The solubility of Au is 7.8% at 220 °C, according to the data in [13].

Figure 5. SEM analysis of pad 5 on ceramic substrate: (a) full pad view, (b) particular of the bottom left area, (c) and (d) intermetallics.

In this technology, the 250 nm gold finish is readily consumed into molten solder and the intermetallic compound starts to form, causing embrittlement of the joint. In this case gold diffuses faster in tin. As the gold diffuses in tin it forms intermetallic and as the concentration of gold increases more Sn is incorporated in the intermetallic and the concentration of Pb will increase forming a pro-eutectic and the resulting structure will be a highly concentrated part of Pb with AuSn intermetallic. The difference in diffusion coefficients of gold into tin and lead is the main mechanics of void formation. This is known as Kirkendall effect and seems to be the cause of the degraded yield and tensile strength of our joints. The photo of pad 5 in fig. 5 shows the voids. It is evident the AuSn phase embedded into the solder made the joint brittle with presence of voids between the intermetallics as can be seen in Figure 5(c) and Figure 5(d). This will happen for concentration of gold going from 3 to 75 wt.%. In order to overcome this problem it would be necessary or to stay under 3 per cent or move to a solder that will form the eutectic 80Au-20Sn where all the above-mentioned problems are not observed. It was not possible to analyze the corresponding pad on silicon because the wire was released and the pad is standing in air. A tentative solution to reduce the gold issue is to keep the time above liquidus at minimum, without affecting the wetting, increasing the cooling rate. Above 4°C/s another problem has been found, the silicon cratering under the silicon pad.

CONCLUSIONS

This work highlights the issues related to the development of a novel compliant contacting technology applicable to wafer level probing and to packaging of integrated circuit. The development of the technology posed several issues to the bonding processes used to build the probing structures. The use of gold finish on small bumps (20-40 μm in diameter), is to be taken into careful consideration. Gold diffuse readily into molten solder and degrade solder mechanical characteristics, even at concentrations considered safe for SMT applications, considerably lowering yield when the bump is subject to mechanical stress, as in the pulling process of this technology. Furthermore, thallium additive used in the gold plating bath, diffuses to the surface forming a passivation layer of thallium oxide that necessary has to be removed prior to bonding. A reducing atmosphere bonding process with Ni under-bump metallurgy is to be preferred and gives best result. Removing the tin-lead soldering process is under investigation using copper and gold pillars to further reduce pitch and avoid the soldering issues. Nevertheless, what is sure that is a must to remove the gold noble metal in small pads dimension due to its not nobleness in flux-less bonding.

REFERENCES

1. M. Balucani, Patent No. WO/2007/104799 (20 September 2007).
2. M. Balucani, P. Nenzi, R. Crescenzi, L. Dolgyi, A. Klyshko, V. Bondarenko, IEEE 3rd Electronic System-Integration Conference, 1-6 (2010).
3. M. Balucani, P. Nenzi, R. Crescenzi, F. Palma, V. Bondarenko, A. Klyshko, *Proc. 7th Intl. Workhop on Nanostructured Materials*, 35-38, (2010).
4. T. Sievert et al., Properties of Lead Free Solders Release 4.0 (available on http://www.boulder.nist.gov/div853/lead_free/solders.html), 2002.
5. J. Glazer, P. Kramer and J.W. Morris Jr., *Circuit World* **18** (4), 41-46 (1993).
6. E. W. Hare, Gold Embrittlement of Solder Joints (available on http://www.semlab.com/goldembrittlementofsolderjoints.html), 2010.
7. E. Mielke, NASA Technical Information Paper No. 008, 1977.
8. M. E. Ferguson, C. D. Fieselman, M. A. Elkins, *IEEE Trans. Compon., Packg., Manuf., Technol.* **C20** (3), 188-193 (1997).
9. Hwang J. S. in *Electronic Packaging and Interconnection Handbook*, 4th ed. C. A. Harper (McGraw-Hill, New York, 2005), 5.61-5.63.
10. D. N. Jacobson, G. Humpston, *Gold Bull.*, **22** (1), 9-18 (1989).
11. W. Lin, Y. C. Lee, *IEEE Trans. Adv. Packg.* **22** (4), 592-601 (1999).
12. C. E. Ho, S. Y. Tsai, C. R. Kao, *IEEE Trans. Adv. Packg.* **24** (2), 493-498 (2001).
13. H. Lee, M Chen, *Mat. Sci. & Eng.* **A333**, 24-34 (2002)

Mater. Res. Soc. Symp. Proc. Vol. 1299 © 2011 Materials Research Society
DOI: 10.1557/opl.2011.532

Giant Piezoresistive Variation of Metal Particles Dispersed in PDMS Matrix

Stefano Stassi[1, 2], Giancarlo Canavese[1], Mariangela Lombardi[1], Andrea Guerriero[1,3] and Candido Fabrizio Pirri[1,3]

[1]Centre for Space Human Robotics, IIT-Italian Institute of Technology, C.so Trento 21, 10129 Torino, Italy

[2] Dept. of Physics, Politecnico di Torino, C.so Duca degli Abruzzi 24, 10129 Torino, Italy

[3] Dept. of Materials Science and Chemical Engineering, Politecnico di Torino, C.so Duca degli Abruzzi 24, 10129 Torino, Italy

ABSTRACT

An Investigation of the piezoresistive response of a metal-polymer composite based on nickel conductive filler in a polydimethylsiloxane (PDMS) insulating matrix for tactile sensor application is presented in this paper. Lacking a mechanical deformation, the prepared composites show no electric conductivity, even though the metal particle content is well above the expected percolation threshold. In contrast, when subjected to uniaxial compression, the electric resistance is strongly reduced. A variation of up to nine orders of magnitude was registered. The thickness of the insulating layer between particles decreases when the sample composite is compressed. Therefore, the electric conduction which is related to a tunneling phenomena, increases exponentially. This behavior is further enhanced by the presence of very sharp nanometric spikes on the particles surface which act as field enhancement factors. In the presented work, the piezoresistive behavior of the composite, the stability in time of the resistance value and the response to several cycles of compression and decompression are evaluated on samples with different physical parameters like nickel content, PDMS copolymer/curing agent ratio and thickness.

INTRODUCTION

Electrical conductive polymers have been studied for 40 years and they continue to stimulate both scientific studies and experimental applications. Two ways can be followed to obtain conductive polymers: producing a polymer that is intrinsically conducting, or adding a conductive filler to an electrically insulating polymeric matrix.

The conductive mechanisms of filled polymeric composites can be divided into two main families. In the former, well known as pressure conductive rubber, the randomly distributed electrical conductive particles are in physical contact with each other when the samples are mechanically loaded [1-3]. Furthermore, it has been demonstrated that changing the strain, the variation of the electrical conduction of the sample can be attributed to the change in the conducting particles contacts which increase when the samples are mechanically deformed [2]. In literature, different percolation models have been proposed to describe the variation in resistivity as a function of filler concentration [4,5]. The percolation models describe the conduction with the presence of electrical paths between two filler particles, but they generally fail below the percolation threshold.

Hybrid piezoresistive polymers of the second group (known as quantum tunnelling composite) differ from the above mentioned materials since the conductive particles are dispersed very close to each other, but are not in physical contact and fully coated by the insulating matrix.

Despite different models have been proposed to explain the conduction mechanism, such as electrical field induced emission [6], Richardson-Schottky transmission types and Pole-Frenkel conduction [4], the tunnelling effect is the well established and widely accepted model used to predict electrical conduction in these piezoresistive polymer composites [4, 7, 8]. The conductive particles are positioned apart from each other thus, in absence of external load, the electrical resistance of the composite material is infinitely large (close to that of the matrix). When a compressive load is applied, a large reduction in resistivity is observed since the insulating layer between the particles decreases and the fillers form a chain of tunnelling paths.

Bloor et al. [7] obtained a variation of twelve orders of magnitude of resistance for a piezoresistive composite based on nickel powders dispersed in polymeric matrix. The presence of nanostructured, extremely sharp tips on the nickel particle surfaces is responsible for the local charge density enhancement that guarantees the extremely large variation of electrical conduction in response to a mechanical strain. The mechanism of conduction results in a field assisted Fowler-Nordheim tunneling, because the charge injected in the composite will reside on the filler, generating very large electric local field on the surface of the tips.

In contrast, the work reported by Abyaneh et al. [8] about silicone filled with smooth zinc particles, shows a considerable lower piezoresponse at the same applied pressure in comparison to composites based on particles with nano-tips. This confirms the essential role in electrical conduction played by the nanomorphology details on the particle surface.

Since piezoresistive polymer composites have an excellent flexibility, resilience and high friction coefficient, they are good candidates as sensitive elements in several applications that require wide deformation range, low power consumption and easy conformability e.g. as flexible force sensors, tactile sensors, industrial grippers, robotic sensing coatings.

In the present work the effects of sample thickness, polymer composition and nickel to PDMS ratio, on the piezoresistive behaviour of PDMS composites based on nanostructured surface nickel particles are finely estimated. Moreover, for the first time the conductivity characterization under dynamic loads and time stability of the conduction under constant strain have been estimated on quantum tunnelling material.

EXPERIMENT

All samples were prepared starting from nickel powder obtained from Vale Inco Ltd. (type 123) and polydimethylsiloxane (PDMS) purchased by Dow Corning Corporation (PDMS KIT: SYLGARD 184). The particle size of the metallic powder was quoted by the producer in the range between 3.5-4.5μm, but FESEM observations revealed a larger range between 3.5 to 7 μm. The polymer was composed by a copolymer part (viscous liquid) and a curing agent one (liquid).

The composites were prepared by initially dispersing the powder in the copolymer PDMS part by gently mixing by hand, in order to prevent the destruction of the tips on the particles surface and to not modify the electric behavior of the composite [7]. Then the PDMS curing agent was added to the viscous mixture and the blend was gently mixed at room temperature obtaining a dark grey paste. After the mixture was degassed for 1 hour and then poured into PMMA molds with different hollow cavity and then cured in an oven at 75°C for two to three hours. The reticulation time varied according to the different compositions of the samples. The samples of composite were then detached from the mold and baked again in the oven at 150°C for a couple of hours to ensure a complete reticulation of the polymer.

Three parameters were varied during the preparation of the composite samples. The PDMS copolymer/curing agent ratio used in this work were 3.33:1 and 10:1, the thickness of the samples 1, 2 and 3 mm and the ratio between nickel and PDMS was modified from 2.5 to 5.5 by weight within steps of 0.5. All combinations of these three parameters were used to create different composites. The surface area of the samples was kept constant to a square of $10\times10mm^2$.

FESEM micrographs were acquired with a Zeiss Supra™ field emission scanning electron microscopy at 5kV operating voltage.

Electrical and mechanical characterizations were performed at room temperature using a Keithley 2635A source meter connected to a home-made sample holder and coupled with a MTS QTest 10 equipped with a load cell of 500 N. The samples were positioned between two Cu plates used as electrodes for applying a voltage in the direction parallel to the applied uniaxial pressure. The voltage was kept fixed and the currents were measured coupling them with the applied load system. All the operations and measurements performed by the whole apparatus were synchronized with the use of a computer.

DISCUSSION

The composites were first characterized with a Field Emission Scanning Electron Microscopy (FESEM) to investigate how the nickel particles were arranged in the polymer matrix. Figure 1(a) shows an induced crack on top of a sample allowing an analysis both of the interior and the surface. The metallic fillers are completely immersed in the PDMS matrix without getting in touch with each other, because every particle is intimately coated with the polymer, even if the nickel content is well above the percolation limit. Hence, no percolation paths are created inside the sample resulting in an insulating electrical behavior. The PDMS filled with the nickel maintains a low stiffness, then under the effect of a compressive force the composite is deformed and the insulating polymer barrier between each particle is reduced proportionally with the applied load. When the thickness of the PDMS layer is reduced enough to ensure a minimum probability of tunneling conductivity between the particles [9], the resistance starts to decrease. This effect grows exponentially with respect to the reduction of the barrier between the nickel particles. Moreover this phenomena is magnify by the presence of very sharp nanometric spikes on the particle surface resulting in a field enhancement factor on the tip of up to 1000 [10].

Electric measurements were performed in order to analyze the piezoresistive behavior of the composites, as function of the nickel content, the PDMS copolymer-curing agent ratio and the thickness, and to check the stability in time of the conductivity and the response to several cycles of compression and decompression. Figure 1(b) shows the difference in the piezoresistive response to a uniaxial applied pressure of PDMS-Ni composites with a thickness of 1, 2 and 3 mm. The graph shows the behavior for samples with a content of nickel powder of 4.5 times the weight with respect to PDMS and a copolymer/curing agent ratio of 10:1. As expected, the variation in resistance is higher for thinner samples. The piezoresistive response as function of the thickness is not linear, as would be expected from the second Ohm's law, but more pronounced and irregular because the deformation of the samples is strongly dependent on the initial condition of shape, dimension and composition. This underlines that physical quantities like electrical resistivity or electric field are not constant through all the samples and explains why all the data in this work are reported in terms of electrical resistance.

Figure 1(c) shows the dependence of the piezoresistive response of the composite on the content of metallic filler inside the PDMS. Data presented in the graph come from samples with nickel to PDMS content from 2.5:1 to 5.5:1 (higher metal contents were difficult to process and they inhibit the reticulating process), 3mm of thickness and 10:1 of PDMS copolymer/curing agent ratio. Increasing the quantity of nickel, the composite becomes more sensitive to compression, starting to conduct at lower applied pressure. Moreover, the variation of resistance from the insulating condition to the maximum conductivity increases considerably by seven orders of magnitude (nine for other samples). All the composite samples have a value of resistance above 10 MΩ when uncompressed, confirming that most of the nickel particles are completely covered by an insulating layer of polymer and there are no conductive paths inside the composite, as observed during FESEM investigation.

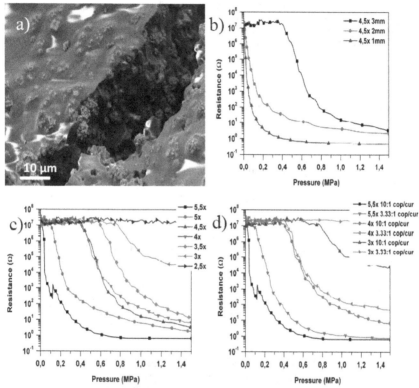

Figure 1. (a) FESEM image of an induced crack in a PDMS-Ni composite, the scale bar corresponds to 10 μm, (b) Thickness dependence of electrical resistance in PDMS-Ni composites, (c) Nickel content dependence of electrical resistance in PDMS-Ni composites, (d) Copolymer/curing agent ratio dependence of electrical resistance in PDMS-Ni composites.

With the same metallic content, larger variations of resistance were measured in samples with a higher copolymer/curing agent ratio. Like PDMS [11], PDMS-Ni composites made with less curing agent do have a lower Young modulus and present less mechanical resistance to compression. Figure 1(d) presents measurements performed on samples with 5.5x, 4x and 3x of

24

the nickel content realized with 3.33:1 and 10:1 of the copolymer/curing agent ratio. All samples with a copolymer/curing agent ratio of 10:1 start to conduct at a lower compressive loading compared to the samples with a ration of 3.33:1 and reach a lower resistance value at the maximum compression.

In order to better understand the sensibility of the material to applied uniaxial pressure, piezoresistive analysis as function of the velocity of applied compression to the samples was performed. As can be seen in Figure 2(a), there is a difference of more than 2 orders of magnitude in the final resistance between pressing at 0.1 mm/min and at 3 mm/min. The response of PDMS-Ni material to a compressive stress is not instantaneous. Inside the polymeric matrix, creep phenomena occur during the compressive test, inducing a delay for the nickel particles to reach the best configuration for tunnel conduction.

The same effect could be seen testing the electrical drift. In Figure 2(b) the values of resistance obtained during compressing the composite sample up to a certain pressure and then keeping it constant for 300 s are reported. It can be seen that the resistance continues to slightly decrease for the first seconds, because the polymer is subjected to a stress relaxation, and then remains constant. Both the graphs in Figure 2 were obtained for a 2 mm sample with a 10:1 copolymer/ curing agent ratio and a content of nickel filler 5 times of the weight with respect to PDMS.

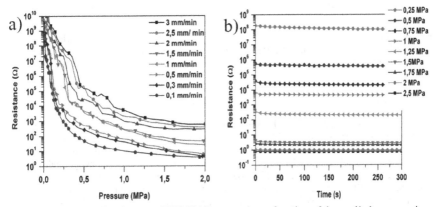

Figure 2. (a) Electrical resistance of PDMS-Ni composite as function of the applied compression rate, (b) Variation in time of resistance of PDMS-Ni composite under different uniaxial applied pressures.

Reusability of the material and repeatability of the measurements are fundamental if the composite should be used for sensor applications. Therefore, we tested how the electrical resistance of the samples varies during ten cycles of compression and decompression from 0 to 2 MPa. Figure 3(a) reports the values of resistance obtained at the maximum compression reached during every cycle for a 2 mm sample with a 10:1 copolymer/curing agent ratio and a content of nickel filler 5 times of the weight with respect to PDMS, while Figure 3(b) shows the complete variation during all the cycles. The elastic behavior of the material allows the composite to completely recover the initial form and particles disposition after every cycle, restoring the resistance value for the insulating state. The maximum compression value shows some variation during the whole measurement, there is a slightly increasing probably due to the hysteresis during the compression and decompression cycle. The four curves in Figure 3(a) evidence how this effect of material creep increases raising the compression velocity and that faster rates result in higher values of resistance when the composite is compressed, as already reported in Figure 2(a).

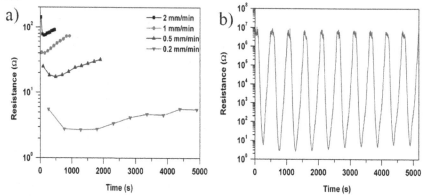

Figure 3. (a) Maximum resistance values of a PDMS-Ni sample obtained for 10 cycles of compression and decompression at different velocity, (b) Variation in time of the electrical resistance of PDMS-Ni composite during 10 cycles of compression and decompression at a velocity of 0.2 mm/min

CONCLUSIONS

This work presents a comprehensive characterization of the piezoresistive response of conductive Ni-PDMS composite. FESEM analyses show that the dispersion recipe preserves the sharp nanostructures tip on the surface of the Ni particles and they also show that the conductive particles are intimately covered by the polymeric matrix. The electrical experiments conducted on undeformed samples indicate that the filler loading does not alter the insulating electrical behavior. The influence of several process parameters (i.e. thickness, composition of the polymer and nickel filler content) have been investigated in terms of variation of electrical resistance. Moreover, the repeatability of the piezoresistive response under a dynamic uniaxial load was reported, as well as the stability of the electrical conduction under different constant compressive loads. The variation of the piezoresistance up to nine orders of magnitude, the simplicity of the process, the encouraging behavior under dynamic load and the time stability make this quantum tunnelling composite an interesting choice for robotic tactile sensor applications.

REFERENCES

1. F. G. Souza, R. C. Michel, and B. G. Soares, *Polymer Testing* **24**, 998-1004 (2005).
2. X. Niu, et al., *Advanced Materials* **19**, 2682-2686 (2007).
3. Q. W. Yuan, et al., *J. Polym. Sci. Part B: Polym. Phys.* **34**(9), 1647-1657 (2006).

4. R. Strumpler and J. Glatz-Reichenbach, *Journal of Electroceramics* **3**(4), 329-346 (1999).
5. F. Carmona, *Physica A: Statistical Mechanics and its Applications* **157**(1), 461-469 (1989).
6. L.K.H. Beek and B.I.C.F. van Pul, *J. Appl. Polym. Sci.* **6**(24), 651-655 (1962).
7. D. Bloor, et al., *J.Phys. D: Appl. Phys.* **38**, 2851-2860 (2005).
8. M.K. Abyaneh and S.K. Kulkarni, *J. Phys. D: Appl. Phys* **41**, 135405 (2008).
9. D. Toker, et al., *Phis. Rev. B* **68**, 041403 (2003).
10. C.J. Edgcombe and U. Valdrè, *Journal of Microscopy* **203**, 188-194 (2001).
11. F.M. Sasoglu, A.J. Bohl and B.E. Layton, *J. Micromech. Microeng.* **17**, 623-632 (2007)

Mater. Res. Soc. Symp. Proc. Vol. 1299 © 2011 Materials Research Society
DOI: 10.1557/opl.2011.396

Characterization of Group III-Nitride Based Surface Acoustic Wave Devices for High Temperature Applications

J. Justice[*], L.E. Rodak[*], V. Narang[†], K. Lee[*], L.A. Hornak[*] and D. Korakakis[*]
[*]Lane Department of Computer Science and Electrical Engineering, West Virginia University, Morgantown, WV 26506 U.S.A.
[†]Department of Physics, West Virginia University, Morgantown, WV 26506 U.S.A.

ABSTRACT

In this study, aluminum nitride (AlN) and gallium nitride (GaN) thin films have been grown via metal organic vapor phase epitaxy (MOVPE) on silicon and sapphire substrates. Samples were annealed at temperatures ranging from 450 to 1000 °C in atmosphere. AlN and GaN thin film quality has been characterized before and after annealing using scanning electron microscopy (SEM), energy dispersive X-ray spectroscopy (EDS) and atomic force microscopy (AFM). Surface acoustic wave (SAW) devices with titanium/platinum interdigital transducers (IDTs) designed to operate at the characteristic frequency and fifth harmonic have been realized using traditional optical photolithographic processes. SAW devices on GaN were thermally cycled from 450 to 850 °C. The S_{21} scattering parameter of SAW devices was measured before and after thermal cycling by a vector network analyzer (VNA). An approach for the suppression of electromagnetic feedthrough (EF) to improve device performance is discussed. Feasibility of 5th harmonic excitation for GHz operation without sub-micron fabrication is also investigated. SAW devices have also been fabricated on the more traditional SAW substrate, lithium niobate ($LiNbO_3$), and device response was compared with those on AlN and GaN at room temperature.

INTRODUCTION

SAW devices are suitable for many different operations such as signal processing, bandpass filtering, pulse compression and sensing. However, most SAW operation and sensing occurs at room temperature. The fundamental operation of SAW devices is due to the piezoelectric effect of the material that the SAW devices are fabricated on. Common piezoelectric materials for SAW devices lose their piezoelectric properties above their Curie temperature, which is typically < 300 °C. AlN and GaN exhibit weaker piezoelectric responses than traditional SAW materials, but retain their piezoelectric properties at higher temperatures. There has been recent research in the development of SAW devices fabricated on AlN that can operate in high temperature environments up to 950 °C [1]. This study focuses on characterizing high temperature SAW response and explores methods for improving the SAW response of devices fabricated on GaN thin films. Those methods include the reduction of EF, which degrades device performance due to the direct coupling of electromagnetic energy between input and output IDTs, and improving the operating frequency by designing SAW devices to operate at the 5th harmonic. 5th harmonic device operation has previously been investigated to increase operating frequency into the gigahertz range without the use of sub-micron fabrication on $LiNbO_3$ [2]. In this study, a similar design has been used to increase the operating frequency of AlN and GaN based SAW devices up to 1.3 GHz with 5 μm IDT finger width and spacing. The acoustic dispersion in GaN thin films was also noticed in this study and compared to other findings from literature.

EXPERIMENT

All materials used in this study were grown in an Aixtron 200/4 RF-S MOVPE system using trimethylgallium (TMGa), trimethylaluminum (TMAl), and ammonia (NH3) as the Ga, Al, and N precursors. Materials used in this work include GaN films on sapphire substrates, AlN films on (100) and (111) silicon substrates, and AlN films grown on 2 μm GaN buffer layers. GaN films and buffer layers were grown using 100 μmol/min of TMGa and a V/III ratio of approximately 700. The reactor temperature was 980 °C and the pressure was 200 mbar. For AlN growths, the TMAl flow rate was 40 μmol/min and the V/III ratio was approximately 1100. The reactor temperature and pressure was 1000 °C and 50 mbar, respectively.

Samples were annealed in atmosphere at temperatures ranging from 450 to 1000 °C in a Thermolyne 6000 series furnace. All samples were annealed for 2 hrs and were gradually cooled to room temperature varying from 6 to 12 hrs. Oxygen content before and after annealing was measured using an INCA x-act ED detector mounted on a JEOL JSM-7600F Field Emission SEM and the INCA Energy software package. Surface roughness before and after annealing was measured using a Digital Instruments MultiMode SPM.

SAW devices used in this study were fabricated using traditional optical photolithographic processes. The S_{21} scattering parameter of SAW devices was measured using an Agilent E8263B vector network analyzer (VNA). Samples were probed using Cascade Microtech ACP40-GSG-150 probes and were connected to the VNA with Cascade Microtech 101-162-B 40 GHz cables. Test setup calibrations to the end of the probe tips were performed using the Cascade Microtech 101-190 impedance standard substrate (ISS). SAW devices were then subjected to thermal cycling from 450 to 850 °C. The S_{21} scattering parameter of SAW devices was measured after each successive annealing.

DISCUSSION

No oxygen in GaN films and less than 3 % oxygen in AlN films was detected by EDS after annealing at 600 °C as can be seen in Figure 1A. AlN and GaN films began to oxidize considerably when annealed at 800 °C. After annealing at 1000 °C, AlN and GaN films were completely oxidized and nitrogen was no longer detectable by EDS (nitrogen scans not shown). Oxidation of AlN and GaN reduce its piezoelectric properties [3], which in turn degrades the overall response of SAW devices.

Surface roughness of AlN films improved slightly with annealing up to 800 °C. Surface roughness of GaN thin films was unchanged after annealing at 600 °C and began to degrade after annealing at 800 °C as shown in Figure 1B and Figure 2. Both AlN and GaN surface roughness degraded rapidly after annealing at 1000 °C. Increased surface roughness results in slower phase velocity and increased insertion loss of SAW devices [4].

SAW devices tested were also fabricated on 3" 41° Y-cut LiNbO₃ wafers. Comparison of device response can be seen in Figure 3A. While GaN films show higher insertion loss than LiNbO₃, the 4.0 μm film showed comparable sidelobe rejection and overall device response. Another consideration in the design of SAW device on GaN is the acoustic dispersion in GaN thin films. Experimentally observed acoustic dispersion in GaN thin films in this study is compared with results from literature, which can be seen in Figure 3B. AlN films were only grown to 700 nm and the device response on these films was relatively poor. It is believed that devices fabricated on thicker AlN layers will show superior responses when compared to device responses on similar thicknesses of GaN layers.

.

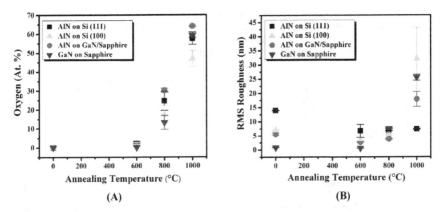

Figure 1. (A) Oxygen content of AlN and GaN thin films as measured by EDS after annealing and (B) surface roughness of AlN and GaN thin films as measured by AFM after annealing. 0 °C refers to as-grown samples prior to annealing.

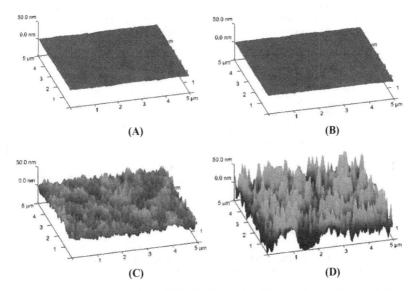

Figure 2. 3-D AFM scans of 4.0 μm GaN thin film surface (A) prior to annealing, and after annealing at (B) 600 °C, (C) 800 °C and (D) 1000 °C.

SAW devices did not show reduced response until annealing at 750 °C as shown in Figure 4, at which point, insertion loss began to increase and the 1st sidelobe rejection began to decrease. After annealing samples at 850 °C, device response was completely lost. There are no measurable data points to include in the plots of Figure 4 at 850 °C.

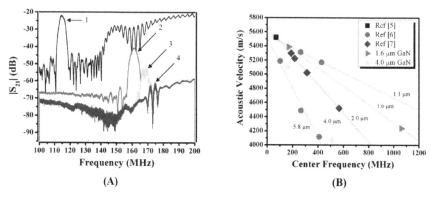

(A)   (B)

Figure 3. (A) Comparison of the response of a SAW device with 8 μm on LiNbO₃ (1), 4.0 μm GaN on sapphire (2), 1.6 μm GaN on sapphire (3) and 700 nm AlN on (111) Si. (B) Experimentally observed acoustic dispersion in GaN on sapphire substrates.

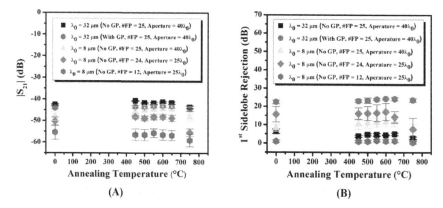

(A)   (B)

Figure 4. (A) Insertion loss at the center frequency as a function of annealing temperature and (B) the 1st sidelobe rejection as a function of annealing temperature.

Effects of EF were drastically reduced with the use of a ground plane, which led to improved sidelobe rejection and a reduction of ripple in the passband as can be seen in Figure 5.

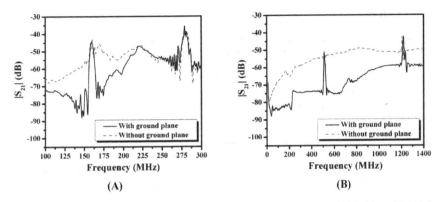

Figure 5. Comparison of device response with and without the use of a ground plane with (A) 8 μm IDT finger width and spacing and (B) 2 μm IDT finger width and spacing. Both devices were fabricated on 4 μm GaN on sapphire.

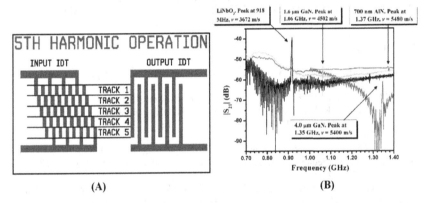

Figure 6. (A) Illustration showing the basic design of a SAW device designed to operate at the 5th harmonic and (B) comparison of devices on LiNbO₃, AlN and GaN operating at the 5th harmonic.

5th harmonic operation was achieved using a similar design to one that has been previously reported [2] and can be seen in figure 6A. Response from devices on LiNbO3, AlN and GaN are shown together in Figure 6B. The 5th harmonic response on the 4.0 μm GaN is at a much higher frequency than expected and is not yet fully understood. The response shown is similar to all responses measured of devices operating at the 5th harmonic on 4.0 μm GaN and an explanation for this result is still being explored.

CONCLUSIONS

GaN thin films were shown to resist oxidation and surface decomposition up to 600 °C. SAW response on GaN did not degrade until thermal cycling at 750 °C. After annealing at 850 °C, SAW response was totally lost. It has been experimentally shown in this study that GaN is an excellent candidate for SAW devices operating up to 600 °C.

AlN films 1 μm and thicker still need to be investigated. It is believed that SAW devices on AlN will have lower insertion loss than devices on GaN and will be able to operate up to 950 °C in atmosphere [1].

Side-lobe rejection of SAW devices was effectively increased with a ground plane, improving the overall response. 5th harmonic operation was shown to be feasible on AlN and GaN thin films without the use of a ground plane. 5th harmonic operation is expected to improve with the use of a ground plane similar to the improvement shown in devices operating at the fundamental frequency.

ACKNOWLEDGEMENTS

This work was supported in part by a grant from the West Virginia Graduate Student Fellowships in Science, Technology, Engineering and Math (STEM) program to LER and AIXTRON. J. Justice would like to thank Vamsi Kumbham for his assistance with e-beam evaporation and rapid thermal annealing used for device fabrication.

REFERENCES

[1]   T. Aubert, O. Elmazria, B. Assouar, L. Bouvot, and M. Oudich, *Appl. Phys. Lett.,* **96**, 203503 (2010).

[2]   P.M. Naraine and C.K. Campbell, *IEEE Ultrason. Symp.*, **0090-5607/84**, 93 (1984).

[3]   R. Farrell, V. R. Pagán, A. Kabulski, S. Kuchibhatla, J. Harman, K. R. Kasarla, L. E. Rodak, P. Famouri, J. Hensel, and D. Korakakis, *Mater. Res. Soc. Symp. Proc.*, **1052**, 1052-DD06-18 (2008).

[4]   Duy-Thach Phana and Gwiy-Sang Chung, *App. Sur. Sci.*, **257(9)**, 4339 (2011).

[5]   S.H. Lee, H.H Jeong, S.B. Bae, H.C Choi, J.H Lee and Y.H Lee, *IEEE Trans. Elec. Dev.*, **48(3)**, 524 (2001).

[6]   S. Petroni, G. Tripoli, C. Combi, B. Vigna, M. De Vittorio, M.T. Todaro, G. Epifani, R. Cingolani and A. Passaseo, *Supperlattices and Microstructures*, **36**, 825 (2004).

[7]   K.H. Choi, H.J. Kim, S.J. Chung, J.Y. Kim, T.K. Lee and Y.J Kim, *J. Mater. Res.*, **18(5)**, 1157 (2003).

Mater. Res. Soc. Symp. Proc. Vol. 1299 © 2011 Materials Research Society
DOI: 10.1557/opl.2011.467

Transport Model for Microfluidic Device for Cell Culture and Tissue Development
Niraj Inamdar[1,2], Linda Griffith[2,3], Jeffrey T. Borenstein[1]

[1]Charles Stark Draper Laboratory, Inc., Department of Biomedical Engineering,
555 Technology Square, Cambridge, MA 02139, U.S.A.
[2]Massachusetts Institute of Technology, Department of Mechanical Engineering,
77 Massachusetts Avenue, Cambridge, MA 02139, U.S.A
[3]Massachusetts Institute of Technology, Department of Biological Engineering,
77 Massachusetts Avenue, Cambridge, MA 02139, U.S.A

ABSTRACT

In recent years, microfluidic devices have emerged as a platform in which to culture tissue for various applications such as drug discovery, toxicity testing, and fundamental investigations of cell-cell interactions. We examine the transport phenomena associated with gradients of soluble factors and oxygen in a microfluidic device for co-culture. This work focuses on emulating conditions known to be important in sustaining a viable culture of cells. Critical parameters include the flow and the resulting shear stresses, the transport of various soluble factors throughout the flow media, and the mechanical arrangement of the cells in the device. Using analytical models derived from first principles, we investigate interactions between flow conditions and transport in a microfluidic device. A particular device of interest is a bilayer configuration in which critical solutes such as oxygen are delivered through the media into one channel, transported across a nanoporous membrane, and consumed by cells cultured in another. The ability to control the flow conditions in this membrane bilayer device to achieve sufficient oxygenation without shear damage is shown to be superior to the case present in a single channel system. Using the results of these analyses, a set of criteria that characterize the geometric and transport properties of a robust microfluidic device are provided.

INTRODUCTION

Microfluidic devices have become a common platform on which biomedical diagnostics can be carried out and cell culture systems observed and engineered [1,2]. More specifically, these microfluidic devices have been applied towards organ transplant and organ assist [3], drug delivery[4], drug discovery [5], and bioassay [6] applications. A common configuration consists of a single microfluidic channel through which culture medium is flowed, with cells cultured on the bottom [7] (Figures 1a, 1c). In general, solute is introduced into the channel, with medium flowed directly over cells, and cellular behavior in response to solute concentration is observed.

As a consequence of fluid being flowed over the cell population, fluidic shear is imparted directly on the cells. However, it is known that cells' metabolic activity is sensitive to shear stresses [8,9,10]. It is very possible, then, that cellular responses that are measured are altered by the presence of shear. Moreover, delivering the same amount of solute to all the cells cultured may become a challenge, for in a single channel configuration, equitable delivery may only occur by increasing the flow rate and consequently, shear. Hence, it is also preferable experimentally to deliver a controlled quantity of solute to the cell population while at the same time having independent control over the imparted shear. We may attempt, then, to seek an alternative device configuration, some of which are analyzed in the literature, and utilize unique biologically-inspired [9] or grooved geometries [10] to shield cells from shear. Here, we will analyze the bilayer construct [1,11,12], which affords itself to a much more straightforward analysis and offers greater simplicity in the way of fabrication.

A bilayer device comprises two channels through which fluid may be flowed, and which are separated by a membrane (Figures 1b,1d). One channel may be used to populate the cell culture (the "cell compartment"), while the other may be used to flow medium containing the solute (the "flow channel"). Solute diffuses across the membrane and into the cell compartment where it is consumed by the cell population. The primary benefit of this construct is that the cells are not subjected to the shear forces of the flow channel, so we may vary that flow rate independently of the cell compartment's to deliver solute to cell compartment. We then only need to maintain a nominal perfusion flow in the cell compartment, while we may modulate the flow channel flow rate arbitrarily.

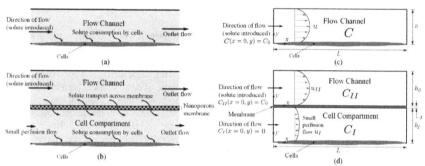

Figure 1. Schematic of (a) single channel and (b) bilayer configurations and geometry of (c) single channel and (d) bilayer configuration

THEORY

In order to establish the fundamental behavior of the configuration, we consider a simple, two-dimensional model with low Reynolds number flow with geometrical values as defined in Figures 1c and 1d. The equations governing transport in the channels are

$$(1) \qquad u_i(y)\frac{\partial C_i}{\partial x} = D_i \frac{\partial^2 C_i}{\partial y^2}, \ i = I, II$$

where $u_i(y)$ is the velocity of the medium in the channel, given by

$$(2) \qquad u_i(y) = \frac{h_i}{2\mu_i}\left(-\frac{\partial p_i}{\partial x}\right)\frac{y_i}{h_i}\left(1 - \frac{y_i}{h_i}\right), \ i = I, II$$

The shear imparted on the channel walls is given by

$$(3) \qquad \tau_i = \mu_i \frac{\partial u_i}{\partial y}\Big|_{y=0,h_i}$$

where μ_i is the viscosity of the medium, and the boundary conditions in the flow channel are

$$(4) \quad \begin{cases} C_{II}(x=0,y)=C_0 \\[2mm] D_{II}\dfrac{\partial C_{II}}{\partial y}\Bigg|_{y=h_{II}} = 0 \\[2mm] D_{II}\dfrac{\partial C_{II}}{\partial y}\Bigg|_{y=0} = \dfrac{D_{membrane}}{t}\big[C_{II}(x,y=0)-C_I(x,y=h_I)\big] \end{cases}$$

with prescribed inlet concentration, no-flux condition at the upper wall, and diffusion across the membrane. Boundary conditions in the cell compartment are

$$(5) \quad \begin{cases} C_I(x=0,y)=0 \\[2mm] D_I\dfrac{\partial C_I}{\partial y}\Bigg|_{y=h_I} = D_{II}\dfrac{\partial C_{II}}{\partial y}\Bigg|_{y=0} \\[2mm] D_I\dfrac{\partial C_I}{\partial y}\Bigg|_{y=0} = V_{max}\rho_{cells}\dfrac{C_I(x,y=0)}{K_M+C_I(x,y=0)} \\[2mm] \qquad\qquad \approx \dfrac{V_{max}\rho_{cells}}{K_M}C_I(x,y=0) \end{cases}$$

assuming no solute at inlet, and cells consuming solute according to first order approximation of Michaelis-Menten kinetics. In order to solve the system given by (1), (3) and (4), a separation of variables technique may be used by writing $C_i = \xi_i(x)\eta_i(y)$ and solving the resulting ordinary differential equations. The differential equation for $\eta_i(y)$ may be solved for by expressing it as a power series in y. Subsequently, the complete solution may be written as

$$(6) \quad C_{II}=C_0\sum_{i=1}^{\infty}A_{II,i}\exp\left(-\frac{\lambda_{II,i}^2 D_{II}}{4u_{0,II}}\frac{x}{h_I^2}\right)\eta_{II,i}(y/h_{II})$$

$$(7) \quad C_I=C_0\sum_{i=1}^{\infty}A_{I,i}\exp\left(-\frac{\lambda_{I,i}^2 D_I}{4u_{0,I}}\frac{x}{h_I^2}\right)\eta_{I,i}(y/h_I)$$

where $\lambda_{j,i}$ $(j=I,II)$ are the zeros of a characteristic polynomial determined by the boundary conditions (4) and (5) and each $\eta_{j,i}$ is characteristic to $\lambda_{j,i}$. The $A_{j,i}$ are determined from an orthogonality condition similar to that used in Fourier analysis, but extended for the case of two adjacent domains. Finally, solute consumption is given by

$$(8) \quad \text{Consumption} = D_I\frac{\partial C_I}{\partial y}\Bigg|_{y=0} = \frac{V_{max}\rho_{cells}}{K_M}C_I(x,y=0)$$

Plots of equations (7) and (8) for the bilayer and the equivalent for the single channel case were generated in MATLAB (see below).

RESULTS AND DISCUSSION

The model system is a hepatocyte culture consuming oxygen. Studies suggest metabolic activity is negatively affected when shear exceeds $\tau_{max} \equiv 1.4\, dynes/cm^2$ [8]; the flow rate Q_{max} associated with τ_{max} is $.72\,\mu L/min$. The hepatocytes are assumed to encounter a hypoxic environment *in vivo*, and the membrane diffusivity chosen is that for a typical polymeric membrane (e.g. polycarbonate or Nafion); all other parameters used are summarized in Table 1. For the bilayer, the concentration field and consumption in the cell compartment may be modulated by increasing the flow rate in the flow channel (Figures 2 and 3), while flow in the cell compartment may be kept at some nominal level. We can impose a more uniform concentration and consumption profile in the cell compartment without increasing Q_I and thus the shear imparted on the cells τ (Figures 2a, 2b, 2c, and 3). Overall transport may be increased by changing the membrane diffusivity, possible, for instance, if the membrane is composed of a crosslinked network (Figure 4).

Parameter	Value	Note
D_{II}, D_I	$3 \times 10^{-5} cm^2/s$	Diffusion coefficient for oxygen in water
$D_{membrane}$	$5 \times 10^{-6} cm^2/s$	Diffusion coefficient for oxygen in membrane
t	$10 \mu m$	Thickness of membrane
μ	$1 \times 10^{-2} dyne \cdot s/cm^2$	Viscosity of water
h_I	$50 \mu m$	Height of channel
w	$200 \mu m$	Width of channel
L	$1.5 cm$	Length of channel
V_{max}	$.5 \times 10^{-9} mol/s/10^6 cells$	Maximum uptake rate of oxygen
K_M	$.5 mmHg = 8.5 \times 10^{-9} mol/cm^3$	Michaelis-Menten parameter for oxygen
C_{sat}	$2.15 \times 10^{-7} mol/cm^3$	Oxygen concentration at saturation
C_0	$2.15 \times 10^{-9} mol/cm^3$	$.01 \times C_{sat}$, 1% saturation
ρ_{cells}	$2.5 \times 10^4 cells/cm^2$	Cell density
Q_{max}	$.72 \mu L/min$	Flow rate; determined from maximum shear condition

Table 1: Summary of parameters used.

For comparison, the single channel profile was also calculated. Even if $Q = Q_{max}$, the concentration profile is highly nonuniform (Figures 2d and 3). Consumption downstream in the single channel case can only match that of the bilayer if flow rates and shear are increased several fold their maximum values (e.g. Figure 3, in which Q is increased to $7Q_{max}$ and τ to $7\tau_{max} = 9.8 \, dynes/cm^2$).

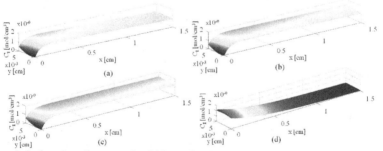

Figure 2. Surface plots of concentration fields as a function of space. For (a), (b) and (c), flow channel rates Q_{II} are $4.5 \, \mu L/min$, $9.0 \, \mu L/min$, and $18.0 \, \mu L/min$, respectively, with cell compartment flow rate $Q_I = Q_{max}/2 = .36 \mu L/min$. For (d), the single channel case, $Q = Q_{max} = .72 \mu L/min$.

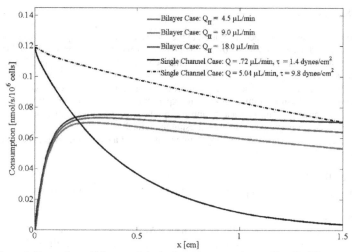

Figure 3: Comparison of bilayer and single channel consumption profiles. For bilayer case, $Q_I = Q_{max}/2 = .36\mu L / min$ (colored lines) with Q_{II} varied. $\tau = .7\ dynes / cm^2$ for each case. For the single channel case, $Q = Q_{max} = .72\mu L / min$ (solid black line) and $Q = 7Q_{max}$ (dotted black line).

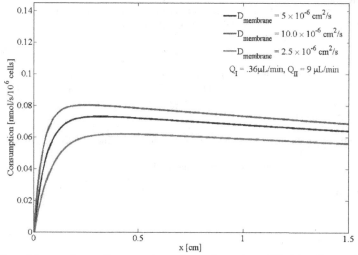

Figure 4. Increase in overall transport due to changes in membrane diffusivity. $D_{membrane}$ is $2.5\times10^{-6}\,cm^2 / s$, $5.0\times10^{-6}\,cm^2 / s$, and $7.5\times10^{-6}\,cm^2 / s$.

CONCLUSIONS

We have considered and developed a solution to the problem of determining the concentration profile in a bilayer device. The bilayer offers a nearly uniform concentration profile, in direct contrast to the nonuniform profile of a standard single-channel device. The bilayer allows this profile to be delivered at a minimal level of shear, while in the single-channel device, transport cannot be made more uniform without increasing shear. It appears that this device configuration can offer a robust platform for future cell-culture experimentation, with modularity and control of solute delivery and microfluidic shear an intrinsic part of its operational capabilities. Validation and extension of the current model using numerical techniques is currently taking place with positive results, and experimental implementation of a cell-cultured bilayer device is under development.

ACKNOWLEDGMENTS

We gratefully acknowledge the support of the NIH NIBIB, grant # 5R01EB010246-02.

REFERENCES

1. A. Carraro, W. Hsu, K.M. Kulig, W.S. Cheung, M.L. Miller, E.J. Weinberg, E.F. Swart, M. Kaazempur-Mofrad, J.T. Borenstein, J.P. Vacanti, and C. Neville, Biomed. Microdevices, **10**, 795–805 (2008).
2. J.T. Borenstein, in *Comprehensive Microsystems*, edited by Y.B. Gianchandani, O. Tabata, and H. Zappe, (Elsevier , Amsterdam, 2005) **2**, pp. 541-584.
3. M.R. Kaazempur-Mofrad, J.P. Vacanti, N.J. Krebs, and J.T. Borenstein, Solid-State Sensor, Actuator and Microsystems Workshop, Hilton Head Island (2004).
4. Z. Chen, S.G. Kujawa, M.J. McKenna, J.O. Fiering, M.J. Mescher, J.T. Borenstein, E.E. Leary Swan, and W.F. Sewell, *J. Controlled Release*, **110**, 1-19 (2005).
5. D.A. LaVan, D.M. Lynn, and R. Langer, *Nature Reviews Drug Discovery*, **1**, 77-84 (2002).
6. S.K. Sia and G.M. Whitesides, *Electrophoresis*, **24**, 3563-3576 (2003).
7. Y. Zeng, T.S. Lee, P. Yu, P. Roy, and H.T. Low, *J. Biomech. Eng.*, **128**, 185-194 (2006)
8. Y. Tanaka, M. Yamato, T. Okano, T. Kitamori, and K. Sato, Meas. Sci. Technol., **17**, 3167-3170 (2006).
9. P.J. Lee, P.J. Hung, and L.P. Lee, *Biotechnol. Bioeng.*, 97, 1340-1346 (2007).
10. J. Park, F. Berthiaume, M. Toner, M. L. Yarmush, A. W. Tilles, *Biotechnol. Bioeng.*, **90**, 632-644 (2005).
11. D.M. Hoganson, J.L. Anderson, E.F. Weinberg, E.J. Swart, B.K. Orrick, J.T. Borenstein, and J.P. Vacanti, *J. Thorac. Cardiovasc. Surg.*, **140**, 990-995 (2010).
12. J.T. Borenstein, *Mater. Res. Soc. Symp. Proc.*, **1139**, 1139-GG02-01 (2008).

Mater. Res. Soc. Symp. Proc. Vol. 1299 © 2011 Materials Research Society
DOI: 10.1557/opl.2011.68

Refractive Index memory effect of ferroelectric materials by domain control

Kazuhiko INOUE and Takeshi MORITA
Graduate School of Frontier Sciences, The University of Tokyo.
5-1-5 Kashiwanoha, Chiba, Kashiwa 277-8563, Japan.

ABSTRACT

Ferroelectric electro-optics materials are widely studied for optical applications, such as optical switches, optical scanners, and optical shutters. However, conventional operation of those devices requires a continuous external electrical field. On the other hand, our group proposes an optical property memory effect by controlling domain structure as either full-polarized or depolarized state using asymmetric voltage operation. The optical property memory effect can keep its optical value, such as refractive index and light transmittance without any external electrical field. In this study, it was confirmed that the refractive index state had two stable values depending on domain conditions. This memory effect should be useful for innovative optical switch or scanner in the future.

INTRODUCTION

Lead lanthanum zirconate titanate (PLZT) is a ferroelectric electro-optic material that has excellent transparency from a visible to an infrared wavelength, and shows variable light transmittance and refractive index with an applied electric voltage[1-4]. Until now, the practical device applications, such as optical switches, optical scanners and optical shutters have been widely studied[5-6]. Such conventional devices require a continuous external electrical field. Alternatively, our group has focused on the memory effects by using an imprint electrical field. An imprint electrical field is observed mainly in ferroelectric thin films and most studies have focused to remove the imprint electrical field[7]. With the imprint electrical field, ferroelectric properties, such as strain, permittivity, light transmittance and refractive index shifts to the direction of the axis of the electric field and ferroelectric materials have two different states at zero electric field. Same to the thin film, bulk ferroelectric materials can also have this imprint electrical field, which is induced by applying a high voltage to ferroelectric material at high temperature. Using this method, ferroelectric properties memory effects such as, strain, permittivity, refractive index and light transmittance had been demonstrated[8-11]. In other words, the devices using memory effect can be driven by pulse voltage operation. However, with those studies, it was clarified that the strain memory gap was reduced by successive pulse operations around 10^4 times[8]. This tendency is expected to be in the PLZT optical memory effect if the imprint electrical field is used for the memory effect. To overcome this fatigue problem, following principle was developed for the memory effects without the imprint electrical field. With this principle, the strain and light transmittance memory effect, which have two stable states at zero electric field were successfully demonstrated[12-13]. This method uses an asymmetric voltage operation to control its domain state. With the shape memory piezoelectric actuator, it was verified that the asymmetry pulsed voltage operation had much better fatigue properties because it is free from the imprint electrical field change[12].

In this study, a refractive index memory effect is demonstrated by controlling the domain condition. This memory enables the realized pulse voltage operation, and can keep the on or off state without any voltage operation. Therefore, energy consumption decreases, and a simple operation becomes possible.

PRINCIPLE

Figure 1 shows the principle of the refractive index memory effect, which is based on the non-linear electro-optic effects. The electro-optic property under bipolar electrical field is shown by dashed lines in the left figure. As shown above, the refractive index doesn't have a memory effect because the refractive index crosses at one state in y-axis. On the other hand, when the amplitude of applied electric field is controlled to induce the minor loop of polarization shown by solid line in right figure, the refractive index obtains two stable states at 0 electric field. One state is polarized state, and other state is depolarized state. These two states corresponds each different refractive index. Thus, the refractive index memory effect can be realized in the ferroelectric electro-optic materials by using the asymmetric voltage operation.

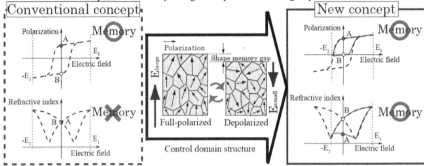

Figure 1. Principle of memory effect with asymmetric voltage operation.

The refractive index memory effect can be controlled using pulse voltage as shown Figure 2. A positive pulse voltage is applied to the PLZT in order to align all polarization. After the positive pulse electric field is removed, the refractive index state goes to ② and is kept without electric field. In order to change the refractive index, a negative pulse voltage is required, which corresponds to the coercive electrical field, to randomize its polarization state. The refractive index state becomes ④ after removal of negative pulse electrical field. Therefore, the optical refractive index memory can be realized as the difference between these two stable refractive index values.

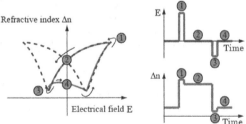

Figure 2. Pulse voltage operation for the refractive index memory effect.

RESULT AND DISCUSSION
Experiment systems

PLZT is an excellent electro-optic material, especially when the lanthanum value is around 8%. In this study, PLZT (La/Zr/Ti=8.19/65/35, $5 \times 5 \times 0.5$ mm^3) plate was used. Two electrodes were attached to its surfaces with conductive tape, as shown in Figure 3. Top electrode was designed to be a prism shape to examine the refractive index change from optical path. Measurement system is shown in Figure 4. When a laser beam penetrates PLZT, it goes straight in the absence of an electric field. On the other hand, applying an electrical field, the optical path is refracted due to the change of the refractive index under the prism shaped top electrode. The laser beam (Edmund 61318-I, 633 nm, 0.8mW, spot size: 0.48 mm) run through PLZT, and the voltage was supplied from the ferroelectric measurement system (Radient Tech.) through a high voltage power supply (NF HVA-4321). A two-dimensional position sensor (Hamamatsu Photonics S1880) was set at 200mm from PLZT to detect the change of the optical path. From the irradiated position change and the distance between PLZT and two-dimensional position sensor, the modification angle θ which is defined in Figure 3 was calculated.

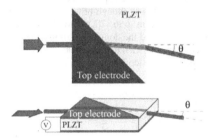

Figure 3. Electrodes of PLZT

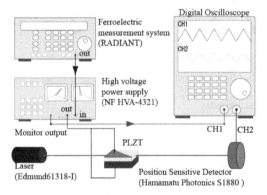

Figure 4. Schematic of measurement setup

Experiment results

The optical path angle change and the polarization as a function of the electrical field are shown in Figure 5. The optical path angle was modified by the refractive index change. When PLZT was applied symmetrical electrical field with the amplitude of 8kV/cm, the polarization inversion was occurred in the both positive and negative direction. At an electric field 0 kV/cm, the domain state was aligned perfectly in each direction. With such a symmetric electric field, the refractive index states are the same, regardless of polarization directions. Therefore, the PLZT didn't have refractive index memory effect. On the other hand, the hysteresis curves under asymmetrical electric field deformed. When the negative amplitude of the applied electric field was much smaller than coercive electric field, the butterfly shaped curve wasn't induced. Figure 6 shows the relationship between negative amplitude and the amount of memory effect which was defined as the gap of refractive index values at zero electrical field. With the amplitude of 8 [kV/cm], the domain state with positive side is aligned because electrical field 8[kV/cm] is enough to be polarized, and the applied electrical field on negative direction was changed from -1[kV/cm] to -7[kV/cm]. The amount of memory effect reached its maximum value around coercive electrical field, -5[kV/cm]. When the applied electrical field was from -1 to -4[kV/cm], some of the domains remained aligned even though other domains were still randomly oriented. While the applied electrical field was from -6 to -7[kV/cm], that exceeded the coercive electrical field, some polarization was reversed, and the domain conditions began to align to one direction. This situation resulted in reducing the memorized value. When the external electrical field corresponded to the coercive field, -5 [kV/cm], the maximum memory value was obtained.

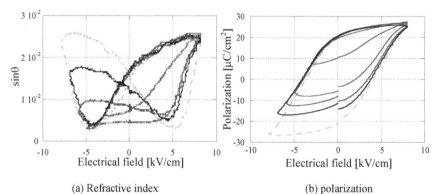

(a) Refractive index   (b) polarization

Figure 5. Hysteresis under the electrical field with various amplitudes

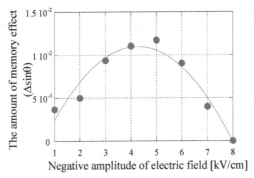

Figure 6. Relationship between the memory effect value and negative amplitude of electric field

Control the optical path using an asymmetric voltage operation

On the basis of these results, the refractive index of PLZT was controlled using a pulse electrical field, as shown Figure 7. The applied electrical field was 8kV/cm in the positive direction, while -5kV/cm in negative direction, and the pulse width was 100ms. Without an external electrical field, the refractive index was remained stable and unchanged in value. In this operation, the amount of memory effect $\Delta\sin\theta$ was 7.4×10^{-3}.

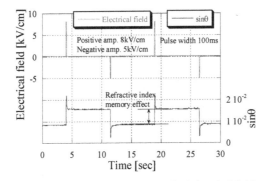

Figure 7. Control of the optical path using asymmetric pulsed electrical field

CONCLUSIONS

In this study, the principle of the refractive index memory effect was proposed. It utilizes the domain condition control by the asymmetric electrical field. By using a ferroelectric electro-optic material PLZT, the asymmetric pulse shaped voltage operation enables the refractive index memory effect. The amount of memory effect was due to domain conditions which were controlled by the asymmetric electrical field operation. To achieve the maximum value of the memory effect, in the positive direction, a sufficient-large voltage is applied to be fully polarized state, and in opposite direction, the amplitude of the voltage is adjusted to the coercive electrical field to be depolarized state. In our next work, the detailed mechanism of memory effect with asymmetrical voltage operation will be investigated to select the optimum materials of optical memory effect.

REFERENCES

1. J. R. Maldonado and A. H. Meitzler, *Proc. of IEEE,* **59**, 368 (1971).
2. W. D. Smith and C. E. Land, *Appl. Phys. Lett.,* **20**, 169 (1972).
3. C. E. Land and W. D. Smith, *Appl. Phys. Lett.,* **23**, 57 (1973).
4. C. E. Land, *Ferroelectrics,* **7**, 45 (1974).
5. K. Uchino, Ceramics International, **21**, 309 (1995).
6. P. E. Shames, P. C. Sun, and Y. Fainman, Appl. Opt. **37**, 3717 (1998).
7. T. Ohashi, H. Hosaka, and T. Morita, *Jpn. J. Appl. Phys.,* **47**, 3985 (2008).
8. T. Ohashi, H. Hosaka, and T. Morita, *Appl. Phys. Lett.,* **93**, 192102 (2008) .
9. T. Morita, Y. Kadota, and H. Hosaka, *Appl. Phys. Lett.,* **90**, 082909 (2007).
10. Y. Kadota, H. Hirota, and T. Morita, *Jpn. J. Appl. Phys.,* **47**, 217 (2008).
11. Y. Kadota, H. Hosaka, and T. Morita, *Proceedings of the Material Research Society Fall meeting 2008*, C9-32, (2008).
12. Y. Kadota, H. Hosaka, and T. Morita, *Proceedings of the JSME-IIP/ISPS,* 331(2009).
13. K. Inoue and T. Morita, *J. Kor. Phys. Soc.,* **57**, 855 (2010).

Mater. Res. Soc. Symp. Proc. Vol. 1299 © 2011 Materials Research Society
DOI: 10.1557/opl.2011.465

Synthesis and Control of ZnS Nanodots and Nanorods with Different Crystalline Structure from an Identical Raw Material Solution and the Excitonic UV Emission

Masato Uehara[1], Satoshi Sasaki[2], Yusuke Nakamura[2], Changi Lee[1], Hiroyuki Nakamura[1] and Hideaki Maeda[1,2,3]
[1]National Institute of Advanced Industrial Science and Technology (AIST), 807-1, Shuku-machi, Tosu, Saga 841-0052, Japan
[2]Kyusyu University, 6-1, Kasuga-kouen, Kasuga, Fukuoka 816-8580, Japan
[3]CREST, Japan Science and Technology Agency, 4-1-8, Hon-cho, Kawaguchi 332-0012, Japan.

ABSTRACT

We synthesized ZnS nanocrystals from identical raw material solution by the thermal decomposition of an amine complex. The shapes of products were changed by simply varying heating rate. At higher heating rate, we obtained the isotropic zincblende nanocrystals. At the lower heating rate, the nanorods were formed and the length was increased with the decrease of heating rate. The nanorods had wurtzite structure below 175 °C, and consequently transformed to zincblende phase during a temperature rise to 200 °C. These particle shapes and phases were related to the adsorption properties of amine ligands. Additionally, the synthesized ZnS nanodots and nanorods exhibited predominantly band-edge emission in fluorescence spectra.

INTRODUCTION

Nanocrystals (NCs) have received much attention because of their attractive properties.[1, 2, 3] The materials properties are strongly influenced by the size and shape. As a result, control of these factors is essential for development and production of NC materials. High-temperature thermal synthesis in organic solvent is a useful technique for the control of NC structure. Surfactant molecules have generally been used. The adsorption properties of surfactants on the particle surface differ with the atomic arrangements of the surface. Consequently, we can control the particle shape by adequate selection of surfactant type and concentration. Many papers describe shape control of NCs using adequate surfactants.[1, 2, 3] However, those surfactant types and quantities have been selected and optimized to obtain one target shape.

ZnS has direct wide band gap energy of ca. 3.7 eV at ambient temperature.[4, 5] This is one of the attractive materials and recently promising as a large variety of devices[6, 7], besides the phosphors.[8] These attractive optical and piezoelectric properties of ZnS are governed by the size and shape as well as other materials. While many papers have described isotropic ZnS NC synthesis,[9] some papers reported that the nanorods were formed in the synthesis using an amine molecule as a surfactant.[10] In generally, the surfactants adsorb to the material source species and particle surface at low temperature, and desorb from them at high temperature. Therefore, using an adequate heat treatment, we might well be able to control the variable shape from one identical starting material solution.

In this study, we synthesized the well-defined ZnS NCs from one raw material with various heating rates in thermal synthesis, and investigated the influence of heating rate on the crystalline structure and morphology with regard to fluorescence properties. Although many papers describe the fluorescence of ZnS NCs, the papers describing the excitonic emission are few and most of them involved trap-state emission.[11] Consequently, production of ZnS NCs exhibiting excitonic emission without tailing would be a worthwhile mission.

EXPERIMENTAL

We used 1-octadecene and oleylamine as a solvent and surfactant, respectively. These reagents were distilled in vacuum before use. All operations were conducted under an argon atmosphere. The ZnI_2 was dissolved in oleylamine (0.3 mol/L) at ambient temperature. Elemental sulfur was dissolved in 1-octadecene (0.3-1.8 mol/L) under 160 °C heating. After cooling, these solutions were mixed at ambient temperature. These mixtures were used as the raw material solution for the ZnS NCs. The ZnS NCs were obtained by heating with various heating rates. The heating rates of 0.6, 1, 2 and 100 °C/min were conducted in an oil bath regulated using a common thermo-controller. Moreover, we obtained a rapid heating rate using a microreactor. In this system, we can raise the temperature to 200 °C in a few seconds. After reaching 200 °C, we held the samples for 1h at this temperature.

The synthesized NCs were characterized by electron microscopy and X-ray diffractometry. We analyzed the X-ray diffraction (XRD) profiles using Rietveld refinement.[12] In addition, we obtained absorption and fluorescence emission spectra for discussion of optical properties. The emission spectra were obtained under the 285nm excitation.

RESULTS AND DISCUSSIONS

Although we used the exact same raw material for all samples, the particle morphology differed markedly with the heating rate. Fig. 1 shows the scanning transmission electron microscope (STEM) images of the present products with variable heating rates. We obtained spherical NCs at the faster heating rate, and the nanorods were formed at the lower rate. In the rapid heated sample, we obtained the uniform spherical NCs of *ca.* 3.9 nm diameter. XRD analysis indicated that this sample had a cubic zincblende (ZB) structure (Fig.2). The 100 °C/min sample also had ZB phase. The mean particle diameter was *ca.* 4 nm, but the homogeneity was inferior to rapid heated sample. At 2, 1 and 0.6 °C/min heating rate, the rod particles were formed. The fraction of isotropic particles decreased with the decrease of the heating rate, and most particles in the 0.6 °C/min sample exhibited a rod-like shape. Their length increased with decrease of heating rate. As described later, the nanorods would elongate at low temperature, resulting the difference of length.

Fig. 2 indicates XRD profiles of rapid heating and 0.6 °C/min samples. In samples, by rapid heating and 100 °C/min heating, the profiles exhibited that they had ZB phase. In samples by slow heating, we observed very weak peaks around 39 ° and 52 ° showing the existence of hexagonal wurtzite (WZ) structure. The intensity of these peaks increased with the decreasing of heating rate. Additionally, the peak around 29 ° was sharper than other peaks. The analysis by Rietveld refinement, reveals that the samples by 0.6, 1 and 2 °C/min heating rate had both ZB and WZ phase, and the sharp peak around 29 ° resulted from the preferred orientation to <111> in ZB and <001> in WZ phase. The fraction of WZ increased concomitantly with the decrease of the heating rate. However, the dominate phase was not WZ rather than ZB even in the 0.6 °C/min sample. The fraction of ZB was 63% and that of WZ was 37 %.

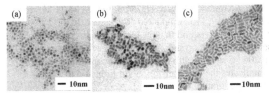

Figure 1. STEM images of the samples synthesized at various heating rates. (a) rapid heating (b) 1°C/min, and (c) 0.6 °C/min.

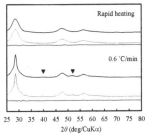

Figure 2. XRD patterns of samples synthesized by rapid heating and 0.6 °C/min heating rates. Thick lines show experimental data. Dotted lines show refined data. Thin lines represent difference values between experimental and refined data.

To clarify the reason for the structural differences among samples, we investigated the temporal evolution of two samples by rapid heating and 0.6 °C/min heating. In rapid heated sample, the temperature reached to 200 °C in a few seconds using the microreactor. As exhibited in Fig.3, the isotropic ZnS NCs had been already formed at 10 s. The crystalline phase was assigned to ZB by high resolution transmission electron microscopy (HR-TEM) and XRD. The morphology of particles was homogeneous; the diameter was *ca.* 2.5 nm. The particles grew to *ca.* 3.9 nm with maintenance of the isotropic shape and ZB crystalline phase during 1 h holding at 200 °C. Thus, we found that in the rapid heating sample, the isotropic NCs which had ZB phase were formed in the early stage, and these NCs grew while keeping the same shape and phase.

In contrast, the temporal evolution of the sample with a 0.6 °C/min heating rate was not as simple. We observed the formation of NCs even at 125 °C. The product was dominated by the elongated shape particles but some isotropic particles were involved. With the temperature rise, the nanorods grew along the long axis up to 175 °C. The mean length was *ca.* 15 nm and the mean diameter of the short axis was almost constant *ca.* 2.5 nm, as shown in Fig. 3. By comparison between Fig.1(c) and Fig. 3 (d), we found that the length of rod particles shrank and the diameter increased around 200 °C. This would be caused by a 1D/2D-ripening in the individual particles, as reported by Peng et al.[13]

The crystalline structure of the sample with a 0.6 °C/min heating rate was changed during heating as well as the particle shape. As shown in Fig.4, we understood that the dominate phases in the 125 and 175 °C samples were WZ. The reflection peak of {002} around 29 ° sharpened and its intensity was increased concomitantly with increasing temperature, indicating the preferred growth along to <001> of WZ during heating. According to Rietveld refinement, the ZB phase was absent below 175 °C. This indicates the transformation from WZ to ZB over

175 °C, because the reaction yield was almost constant 80 % over 175 °C. In the HR-TEM, we confirmed the difference of lattice structure between 175 °C and 200 °C (Fig. 5). In the 175 °C sample, we observed the WZ arrangements, and understood that the elongation direction was <001> in WZ. In contrast, the 200 °C sample had both crystalline phases in each particle. The edge regions of rods had ZB structure in 200 °C sample, although the middle region had WZ structure as well as 175 °C sample. This indicates that the transformation from WZ to ZB occurred at the edges rather than in the middle region of the rods.

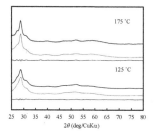

Figure 3. STEM images of the samples. (a) 10 sec-heated sample (rapid heating). (b) 150 °C and (c) 175 °C in the 0.6 °C/min sample.

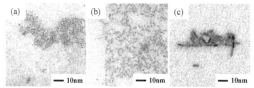

Figure 4. XRD patterns of the sample with a 0.6 °C/min. Thick lines show experimental data. Dotted lines show refined data. Thin lines portray the difference value between experimental and refined data.

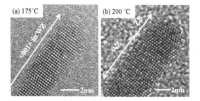

Figure 5. HR-TEM images of the sample with a 0.6 °C/min. The closed packed planes were lying perpendicular to the long axis.

Thus, we found the structural evolution of the sample with a 0.6 °C/min. The particles elongated to <001> of WZ phase below 175 °C. Around 200 °C, the particles shrank and transformed to ZB. The anisotropic growth would be the result of the adsorption property of amine ligands, as some groups reported. According to them, the amine ligands preferentially adsorb to a specific crystallographic face such as {100} and {110} faces of WZ, leading to 1-D growth in ZnS, ZnSe and CdSe NC materials. [10, 14] Around 200 °C, the preferential adsorption would disappear, because the rapid heated sample had isotropic shape. This desorption around 200 °C would induce the change in morphology from anisotropic to isotropic shape, resulting in the shrinking and thickening around 200 °C.

In addition, the transformation from WZ to ZB in this work is strange. In the bulk material, WZ is the higher temperature phase and ZB phase is the lower temperature phase. The present transformation is exactly opposite to thermodynamically transformation in bulk. This transformation could be attributable to the change in surface energy. At the nanoscale, the appearance of WZ was reported by some groups.[15, 16] According to them, this would be caused by the lower surface energy of WZ phase compared with that of ZB phase. The nanomaterials have significant higher surface area, and strongly affects from surface condition. Then, the high temperature phase would appear at ambient temperature. In this work, the surface condition would change around 200 °C. We considered that the transformation to ZB is related to the change of surface condition.

Finally, we examine the fluorescence properties of present products by variable heating rates. Fig.6 illustrates the ultraviolet-visible (UV-vis) absorption and fluorescence (FL) emission spectra of them after 200 °C-1h holding. In the absorption spectra, all products have a broad peak 3.9-4.1eV. These were higher than the ZnS bulk band gap (3.7 eV).[4, 5] This would be caused by the quantum effect according to Transmission Electron Microscopy (TEM) analysis. The peak energy of the 0.6 °C/min sample was smaller than that of the rapid heated sample. This would be attributed to the difference of shape.[17] In the FL spectra, all spectra were dominated by a sharp peak 3.8-4.0 eV. The full width at half maximum (FWHM) of these peaks was narrow. The emission spectra did not change with excitation wavelength of 260-285 nm, and then the sharp peak is attributable to the excitonic emission. The peak position of the FL was shifted with the absorption edge. Thus, we were able to obtain the ZnS NC materials with predominant excitonic emission. Some reports described the excitonic emission of ZnS NCs but the spectra had the tailing or considerable trap site emission by the shallow donor-acceptor emission or deep trap emission.[8, 14] To our knowledge, the papers revealing the narrow excitonic emission without a defect emission are few except as Geng et al.[8] They exhibited a well-defined and sharp excitonic emission from a ZnS nanorod. Although our products exhibited broad peaks around 3.2 or 3.45 eV only to a slight degree, the spectra were almost dominated to the excitonic emission which was similar with Geng et al. In addition, we successfully controlled-synthesized ZnS spherical nanoparticles and nanorods which exhibited the excitonic emission.

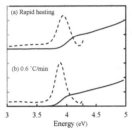

Figure 6. Absorption and fluorescence spectra of products after 200°C heating. (a) rapid heated sample and (b) 0.6 °C/min heating rate sample. Solid and dashed lines represent the absorption and fluorescence spectra, respectively.

CONCLUSIONS

We synthesized ZnS nanocrystals using an identical raw material solution with variable heating rates. Thermal decomposition of amine complex in organic solvent was applied as a synthesis method. We were able to controlled-synthesize the ZnS nanodots and nanorods by the simple varying heating rate. At the higher heating rates, the particle shape was spherical and the dominant phase was ZB. At the lower heating rates, nanorods were formed and the length was increased concomitantly with the decrease of the heating rate. However, the dominant phase in these lower heating samples was also isotropic ZB rather than anisotropic WZ. The WZ nanorods which grew at low temperature transformed to ZB phase around 200 °C. This is opposite to the thermodynamic transformation of the bulk material. The particle shape and crystalline phase seems to be influenced to the adsorption property of amine ligands according to our structural analysis for the temporal evolution of samples.

Additionally, the synthesized ZnS nanodots and nanorods predominantly exhibited band edge emission in fluorescence spectra. The fluorescence emission peak was narrow and had only a slight tail.

ACKNOWLEDGMENTS

A part of this study was supported by an Industrial Technology Research Grant Program (06A21201d) from the New Energy and Industrial Technology Development Organization (NEDO) of Japan. In addition, TEM analysis was supported by the Nanotechnology Network Project (Kyushu-area Nanotechnology Network) of the Ministry of Education, Culture, Sports, Science and Technology (MEXT), Japan.

REFERENCES

1. Y. Jun, J. Choi and J. Cheon, *Angew. Chem. Int. Ed.*, **45**, 3414-3439 (2006).
2. J. Park, J. Joo, S. G. Kwon, Y. Jang and T. Hyeon, *Angew. Chem. Int. Ed.*, **46**, 4630-4660 (2007).
3. P. S. Nair, K. P. Fritz and G. D. Scholes, *Small*, **3**, 481-487 (2007).
4. T. K. Tran, W. Park, W. Tong, M. M. Kyi, B. K. Wagner and C. J. Summers, *J. Appl. Phys.*, **81**, 2803-2809 (1997).
5. Q. Xiong, G. Chen, J. D. Acord, X. Liu, J. J. Zengel, H. R. Gutierrez, J. M. Redwing, L. C. Lew Yan Voon and B. Lassen, P. C. Eklund, *Nano Lett.*, **4**, 1663-1668 (2004).
6. J. A. Zapien, Y. Jiang, X. M. Meng, W. Chen, F. C. K. Au, Y. Lifshitz and S. T. Lee, *Appl. Phys. Lett.*, **84**, 1189-1191 (2004).
7. D. Moore and Z. L. Wang, *J. Mater. Chem.*, **16**, 3898-3905 (2006).
8. P. Yang, M. K. Lu, D. Xu, D. L. Yuan and G. J. Zhou, *J. Lumin.*, **93**, 101-105 (2001).
9. L. S. Li, N. Pradhan, Y. Wang and X. Peng, *Nano Lett.*, **4**, 2261-2264 (2004).
10. B. Y. Geng, X. W. Liu, Q. B. Du and L. D. Zhang, *Appl. Phys. Lett.*, **90**, 183106 (2007).
11. L. S. Li, N. Pradhan, Y. Wang and X. Peng, *Nano Lett.*, **4**, 2261-2264 (2004).
12. F. Izumi and K. Momma, *Solid State Phenom.*, **130**, 15-20 (2007).
13. Z. A. Peng and X. G. Peng, *J. Am. Chem. Soc.*, **123**, 1389-1395 (2001).
14. J. H. Yu, J. Joo, H. M. Park, S. Baik, Y. W. Kim, S. C. Kim and T. Hyeon, *J. Am. Chem. Soc.*, **127**, 5662-5670 (2005).
15. P. Li, L. Wang, L. Wang and Y. Li, *Chem. Eur. J.*, **14**, 5951-5956 (2008).
16. Y. Li, X. Li, C. Yang and Y. Li, *J. Phys. Chem. B*, **108**, 16002-16011 (2004).
17. J. Li and L. W. Wang, *Phys. Rev. B*, **72**, 125325 (2005).

Mater. Res. Soc. Symp. Proc. Vol. 1299 © 2011 Materials Research Society
DOI: 10.1557/opl.2011.254

Improving PZT Thin Film Texture Through Pt Metallization and Seed Layers

L.M. Sanchez[1,3], D.M. Potrepka[1], G.R. Fox[2], I. Takeuchi[3], R.G. Polcawich[1]

[1]U.S. Army Research Laboratory, 2800 Powder Mill Road, Adelphi, Maryland 20783

[2]Fox Materials Consulting LLC, 7145 Baker Road, Colorado Spring, CO 80908

[3]Department of Materials Science and Engineering, University of Maryland, College Park, Maryland 20742, USA

ABSTRACT

Leveraging past research activities in orientation control of lead zirconate titanate (PZT) thin films [1,2], this work attempts to optimize those research results using the fabrication equipment at the U.S. Army Research Laboratory so as to achieve a high degree of {001}-texture and improved piezoelectric properties. Initial experiments examined the influence of Ti/Pt and TiO_2/Pt thins films used as the base-electrode for chemical solution deposition PZT thin film growth. In all cases, the starting silicon substrates used a 500 nm thermally grown silicon dioxide. The Pt films were sputter deposited onto highly textured titanium dioxide films grown by a thermal oxidation process of a sputtered Ti film [3]. The second objective targeted was to achieve highly {001}-textured PZT using a seed layer of $PbTiO_3$ (PT). A comparative study was performed between Ti/Pt and TiO_2/Pt bottom electrodes. The results indicate that the use of a highly oriented TiO_2 led to highly {111}-textured Pt, which in turn improved both the PT and PZT orientations. Both PZT (52/48) and (45/55) thin films with and without PT seed layers were deposited and examined via x-ray diffraction methods (XRD) as a function of annealing temperature. As expected, the PT seed layer provides significant improvement in the PZT {001}-texture while suppressing the {111}-texture of the PZT. Improvements in the Lotgering factor (f) were observed upon comparison of the original Ti/Pt/PZT process (f=0.66) with samples using the PT seed layer as a template, Ti/Pt/PT/PZT (f=0.87), and with films deposited onto the improved Pt electrodes, TiO_2/Pt/PT/PZT (f=0.96).

INTRODUCTION

Lead zirconium titanate (PZT) exhibits superior piezoelectric properties for many types of microelectromechanical systems (MEMS) and is one of the most economical piezoelectric ceramics, making it cost effective for mass production. It also exhibits a high piezoelectric coefficient [4] compared with other common piezoelectric materials such as ZnO and AlN, thus allowing for the use of lower operating voltages while still achieving the same actuator performance metrics. The highest magnitude piezoelectric coefficients are observed at the PZT morphotropic phase boundary, where the crystal structure changes abruptly between the tetragonal and rhombohedral symmetry [5]. The morphotropic phase boundary (MPB) is located around the $PbZr_{0.52}Ti_{0.48}O_3$, or PZT (52/48), composition. At the MPB, high dielectric permittivity and piezoelectric coefficient are observed.

At the Army Research Laboratory (ARL) in Adelphi MD, efforts are focused on achieving highly {001}-textured PZT (52/48). Proper control of the crystalline texture will allow a 30-50% increase in the piezoelectric stress constant. These improvements will result in substantial improvements in device performance including lower actuation voltages, higher force actuation, and lower power consumption.

EXPERIMENT

The starting substrate for all experiments was a 4 in diameter, (100) silicon (Si) wafer coated with a thermally oxidized silicon dioxide (SiO_2) thin film with a thickness of 500 nm. The electrode layer, onto which the PZT thin films were deposited, was sputter-deposited onto the silicon dioxide. The process is detailed in Potrepka et al. [3] The process began with the sputter deposition of a thin film of titanium (Ti) using a Unaxis Clusterline 200 (CLC) deposition tool. After deposition, an oxygen anneal at 750°C in a Bruce Technologies tube furnace was performed to convert the Ti into TiO_2. The TiO_2 acts as a textured seed layer for {111} platinum (Pt) nucleation. A 100 nm Pt film was subsequently deposited at 500°C using the CLC. The resulting highly {111}-textured Pt provides a template for {111}-textured PZT thin film growth. In order to achieve {001}-textured PZT growth, a seed layer deposited on top of the Pt was employed.

Chemical solution deposition (CSD) of PZT allows for control of stoichiometry, reduced processing temperatures, and is relatively cost effective for development and mass production. Based on the work of Muralt et al. [6], the use of $PbTiO_3$ as a seed layer provides a good lattice match with PZT films and provides a template for growth of (100) and (001) perovskite crystals. A single layer of 30% Pb-excess $PbTiO_3$ was deposited via CSD using a manual spinner with a speed of 2500 rpm for 30 sec and then pyrolyzed at 350°C for 2 minutes on a hot plate. The sample was annealed in a Heatpulse 610 rapid thermal anneal (RTA) oven at 700°C for 60 seconds with a temperature ramp of 4°C/sec.

In this research, the PZT solutions were prepared in a procedure modified from that of Budd et al. [7] and varied using 8, 10, and 15% Pb excess to determine the best parameters for {001}-textured film growth. Lead (III) Acetate Trihydrate (99.995%) from Puratronic was mixed with 2-Methoxyethanol (2-MOE) from Sigma Aldrich and refluxed at 120°C for 20 minutes in flowing nitrogen (N_2) using a Heidolp Laborota 4003 control rotary evaporator. Zirconium (IV) n-propoxide (70 wt% in n-propanol), Titanium (IV) Isoproxide (97%), both from Alfa Aesar, and 2-MOE were combined and stirred at room temperature on a magnetic stirrer. All solution preparation apparatus was located in a vented glove-box.

The lead solution that was on the rotary evaporator was then dehydrated by decreasing the pressure (~350 mbar) and allowing the solution to boil resulting in removal of the water and methoxyethoxide [8]. White foam appears demonstrating the end of the dehydration process. The Pb solution was removed from the evaporator and mixed with the Zr/Ti solution (targeted at 52/48 or 45/55) that was mixing on the magnetic stirrer. This new solution was returned to the evaporator for a 3 hour reflux with flowing N_2. A 5 minute dehydration step was performed at 925 mbar followed by a 5 minute N_2 purge. The solution was allowed to cool before adding 4% formamide, from Sigma Aldrich, to the resulting PZT solution volume and finally the solution was stirred overnight inside the glovebox.

The solution was tested by preparing a 500 nm thick PZT thin film on a 2.5 cm by 2.5 cm platinized Si sample and creating 500 μm by 500 μm capacitors with Pt top electrodes. Electrical characterization of the test capacitors yielded a capacitance of ~ 7.2 nF (at 10 kHz), dielectric loss of ~0.03 (at 10 kHz), and $+P_r$ ~15μC/cm^2 and $-P_r$ ~ -13 μC/cm^2 determined from a Radiant Technologies RT-66i measurement unit.

Several experiments were performed on the various PZT solutions that were synthesized. The objective of the first series of tests was to observe the effects of the annealing ramp rate and final temperature on the PZT crystallinity. The RTA temperature for the 400 nm PZT films was varied by 20°C from 600°C-740°C. Below 680°C, the pyrochlore phase is visible and is studied in other reviews [9]. Above 720°C the (110) and (111) PZT peaks begin to dominate the PZT nucleation. For this reason, only the results from samples annealed at 700°C are presented in detail. The second series of tests involved looking into the effect of Pb-excess PZT solutions and the effects of varying the RTA ramp rates. PZT films deposited from solutions with 8, 10, and 15% Pb excess were analyzed following the various RTA ramp processes. The PZT films were deposited onto untreated Pt substrates as well as those coated with a $PbTiO_3$ seed layer. In both test series, an older bottom electrode metal process, namely, Ti (20 nm) / Pt (82 nm) sputter deposited at a substrate temperature of 500°C and a new bottom electrode, namely, TiO_2 (30 nm) / Pt (100 nm) were compared (table 1).

Diffraction measurements were performed using a Rigaku Ultima III diffractometer with Bragg-Brentano geometry. The crystallinity of all the samples was analyzed and noticeable improvements stemming from the optimization of the $PbTiO_3$ seed layer and bottom Pt were observed.

Table 1. Samples made to observe the effects of RTA temperature on PZT nucleation using different bottom electrodes.

Bottom Metal	Seed Layer	PZT (52/48)	PZT (45/55)
Ti/Pt	No	Yes	No
		No	Yes
	Yes	Yes	No
		No	Yes
TiO_2/Pt	No	Yes	No
		No	Yes
	Yes	Yes	No
		No	Yes

RESULTS

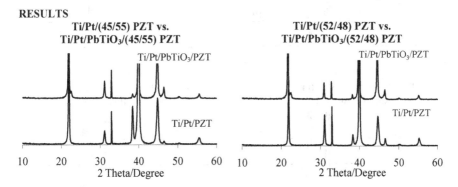

Ti/Pt/(45/55) PZT vs. Ti/Pt/PbTiO₃/(45/55) PZT

Ti/Pt/(52/48) PZT vs. Ti/Pt/PbTiO₃/(52/48) PZT

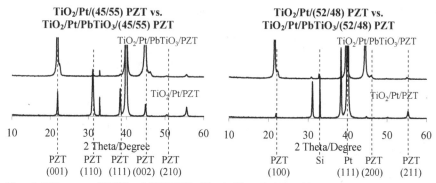

Figure 1. XRD patterns of (45/55) and (52/48) PZT with and without PbTiO3 seed layer. Labels are included at the bottom of the graph set for peak identification.

In (45/55) and (52/48) PZT compositions, samples with the $PbTiO_3$ seed layer (Figure 1) exhibited peak splitting of the (001)/(100) and (002)/(200) peaks as well as a reduction in the 111 relative peak intensity. The (110) PZT peak remains relatively unaffected except in the samples with both the TiO_2/Pt bottom electrode and the $PbTiO_3$ seed layer. The combination of both suppresses the PZT randomly oriented grains and allows for a stronger (001) peak. The next set of tests continued with TiO_2/Pt/$PbTiO_3$/PZT (52/48) since it showed the greatest improvement in {001}-textured PZT according to the XRD data.

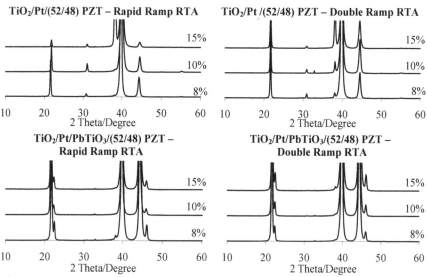

Figure 2. XRD patterns of PZT (52/48). The effects of different ramp rates are compared with varying Pb-excess solutions.

PZT (52/48) thin films with varying Pb-excess solution compositions were analyzed (Figure 2). Solution Pb-excesses of 8, 10, and 15% were studied to determine the best solution composition for {001}-textured PZT. The effects of the RTA ramp were also observed by performing "rapid ramp" and "double ramp" techniques. In a rapid ramp RTA, the sample temperatures were ramped up and down as fast as possible. In the case studied, the temperature was ramped up at 199°C/sec, held at 700°C for 60 seconds and ramped down as fast as the RTA tool could cool. By using such an anneal profile the PZT is expected to crystallize in a textured state with a minimization of composition inhomogeneities created by diffusion and PbO evaporation from the surface of the film.

Similarly, in the double ramp technique, the sample temperature is ramped up at a rate of 199°C/sec until 550°C, stabilized, and held for 2 minutes. The anneal at 550°C allows the PZT to crystallize at low temperature under a condition that results in reduced PbO evaporation from the sample surface but still provides the definition of the crystalline texture. The subsequent high temperature anneal, for which the PZT is ramped up at 199°C/sec to 700°C and held for 30 seconds, allows for grain growth and removal at other growth defects.

$$f = \frac{P - P_0}{1 - P_0}$$

P_0 = XRD intensity values based on a standard
P = XRD intensity values based on actual sample
$I_{(001/100)}$ = Intensity of (001) and/or (100) peaks

$$P = \frac{[I_{(001)} + I_{(002)}]}{\sum I_{(hkl)}}$$

$\sum I_{(hkl)}$ = Sum of all PZT peaks

Equation 1. Lotgering factor equation for (001) orientation is a simple quantification of the degree of orientation of the material.

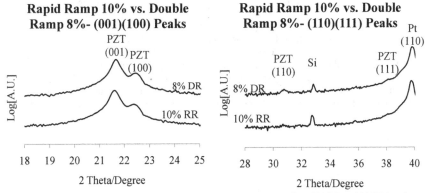

Figure 3. Comparison of the 10% Pb excess RR and 8% Pb excess DR XRD peaks at 001/100 and 110/111 peaks.

By performing a rapid ramp and a double ramp anneal on samples with the optimized bottom electrode and PbTiO$_3$ seed layer, it was determined that the 10% Pb excess with rapid ramp (RR) RTA and 8% Pb excess with double ramp (DR) RTA showed the highest (001) orientation and the greatest (110) and (111) PZT peak reduction (figure 3). Lotgering factors (Equation 1) of the RR and DR were $f_{(001)}$=94.8% and $f_{(001)}$=95%, respectively. Future tests involve performing electrical measurements and determining the capacitance, loss, and dielectric coefficient of the PZT.

CONCLUSION

The XRD data indicates an improvement in the bottom Pt quality allowing for proper nucleation of the PZT film. Through the RTA temperature tests it was determined that by using a 700°C RTA we achieve the best PZT orientations when performed on samples with the TiO_2/Pt electrode and the $PbTiO_3$ seed layer.

The use of the fast ramp rates does not allow sufficient time for complete transformation of the PZT film. Instead, surface nucleation processes dominate film texture. The $PbTiO_3$ allows the PZT film to nucleate along the (001) direction which propagates through subsequent PZT layers. The improved Pt limits the nucleation of (110) and (111) PZT thus allowing even greater 001/100 orientation in the film when combined with the $PbTiO_3$ seed layer.

REFERENCES:

1. N. Ledermann, et.al. "Textured, piezoelectric $Pb(Zr_x, Ti_{1-x})O_3$ thin films for MEMS: intergration, deposition, and properties." Sensors and Actuators A 105 (2003): 162.
2. Trolier-McKinstry, S. "Improved Thin Film Piezoelectrics for Actuator Applications." ARO Final Report. 2007.
3. Potrepka, D. M. "Pt/TiO_2 Growth Templates for Enhanced PZT films and MEMS Devices." MRS Proceedings Session S. Boston, 2010.
4. R.G. Polcawich, D. Judy, J. Pulskamp, S. Trolier-McKinstry, & M. Dubey. "Surface Micromachined Microelectromechancial Ohmic Series Switch Using Thin-Film Piezoelectric Actuators." IEEE (2007): 2083-2086.
5. B. Jaffe, R.S. Roth, & S. Marzullo. "Piezoelectric Properties of Lead Zirconate- Lead Titanate Solid Solution Ceramics." Journal of Applied Physics (1954): 809-810.
6. P. Muralt, T. Maeder, et. al. "Texture Control of $PbTiO_3$ and $Pb(Zr,Ti)O_2$ Thin Films with TiO_2 Seeding." Journal of Applied Physics (1998): 81-89.
7. K.D. Budd, S.K. Dey, & D.A. Payne. "Sol-Gel Processing of $PbTiO_3$, $PbZrO_3$, PZT, and PLZT Thin Films." Electrical Ceramics (1985).
8. B. Kosec, M. Malic, & M. Mandeljc. "Chemical Solution Deposition of PZT Thin Films for Microelectronics." Materials Science in Semiconductor Processing (2003): 97-103.
9. Desu, C.K. Kwok and S.B. "Pyrochlore to Perovskite Phase Transformation in Sol-Gel Derived Lead-Zirconate-Titanate Thin Films." Applied Physics Letters (1992): 1430-1432.

Process Integration

Mater. Res. Soc. Symp. Proc. Vol. 1299 © 2011 Materials Research Society
DOI: 10.1557/opl.2011.52

Patterning Nanomaterials on Fragile Micromachined Structures using Electron Beam Lithography

Kaushik Das, Pascal Hubert, and Srikar Vengallatore

Department of Mechanical Engineering, McGill University, Montreal, QC, H3A 2K6, Canada

ABSTRACT

Integration of nanomaterials (in the form of quantum dots, nanotubes, nanowires, nanocrystalline thin films, and nanocomposite films) with micromachined structures and devices has the potential to enable the development of microelectromechanical systems (MEMS) with enhanced functionality and improved performance. Here, we present a fabrication approach that combines spray-coating of electron beam resist with direct-write electron beam lithography to pattern nanomaterials on fragile micromachined components. Polymers and metallic structures in the form of arrays of holes, concentric circles, and arrays of lines, with critical dimensions ranging from 135 nm to 500 nm, were patterned directly on various micromachined structures including commercial metal-coated silicon microcantilevers used for atomic force microscopy, and commercial plate-mode SiC/AlN microresonators used for sensing.

INTRODUCTION

During the last two decades, there has been tremendous progress in the synthesis and characterization of a wide variety of nanomaterials [1]. Nanoparticles, nanowires, nanotubes, nanocrystalline thin films, and nanocomposite thin films can be fabricated using organic synthesis, self-assembly, nanofabrication, or a combination of these methods. Nanomaterials are excellent candidates for fundamental studies of materials science because many of their physical and chemical properties are significantly different from those exhibited by their bulk counterparts [1]. In turn, understanding the effects of size on process-structure-property relationships can open up new opportunities for exploiting nanomaterials in engineering applications. These opportunities for scientific studies and technological applications are the primary motivation for efforts to integrate nanomaterials with microelectromechanical systems (MEMS). These miniaturized systems are used for a diverse array of applications ranging from sensing and displays to portable power generation and medicine. The performance and functionality of many MEMS can be enhanced by integrating nanomaterials on the micromachined structural components of these systems. The amount of nanomaterials required per device is small; therefore, the lack of availability of high-quality nanomaterials in large quantities is not a barrier. For example, a single carbon nanotube can make a significant difference to the performance of silicon microcantilevers used in scanning probe microscopy. In addition, MEMS-based test platforms are ideally suited for measuring the properties of nanoscale structures with high resolution and accuracy [2, 3]. These measurements can establish a solid foundation for understanding the mechanisms responsible for the size-dependent properties of nanomaterials.

The advantages of integrating nanomaterials with MEMS have been noted before, and several groups have developed different approaches for integrating the synthesis of nanomaterials with microfabrication process flows (see, for example, [4–7]). These approaches can be classified based on two factors. The first is the selection of a method for synthesizing the nanomaterial. This method can be *bottom-up* (using self-assembly and organic synthesis) or *top-down* (using directed patterning, deposition and etching). Self-assembly and organic growth are simple and convenient, but offer relatively little control on location, alignment, and dispersion. Moreover, self-assembled structures are susceptible to large variability from device to device [1]. In contrast, directed patterning involves controlled high energy ultra-violet, x-ray, electron and ion beams to create nanometer-scale patterns with precise control over position, distribution and orientation of nanomaterials [1]. However, many of these techniques are constrained either by high cost or slow patterning speeds, especially for structures that have critical feature sizes less than 100 nm. In some cases, the nanomaterial can be synthesized using hybrid techniques that combine the characteristics of both self-assembly and directed growth.

The second factor is the sequence of nanomaterial synthesis and micromachining. This selection is based primarily on the ability of the nanomaterial to withstand the harsh conditions that are encountered while micromachining the device. These conditions can include high temperatures, which can lead to undesirable grain growth or coarsening, and corrosive environments created by the chemicals used for etching. If the nanomaterial can indeed be shielded from these harsh environments, then this material can be first synthesized at the wafer level (using techniques such as nanoimprint lithography, dip-pen nanolithography and electron beam patterning), and the micromachined structures can then be created [7].

In this paper, we consider the alternate approach wherein the micromachining steps are completed before integrating the nanomaterial. The advantage of this approach is that the steps used for micromachining are not constrained by thermal or chemical compatibility with the nanomaterial. However, this approach must also confront the difficult challenge of processing the nanomaterial on fragile micromachined structures. Thus, any microfabrication process that generates contact forces will generally damage the micromachined components. For example, dispensing photoresist or electron beam resist by spin coating on freestanding micromachined membranes, plates or beams can cause fracture. To avoid these problems we have developed an approach that combines spray-coating of electron beam resist, physical vapor deposition of thin films, and electron beam lithography. The details of these processes, and the fabrication of nanostructures on commercial micromachined plates and beams, are described in this paper.

The next section considers the integration of polymeric nanostructures on silicon carbide plate resonators. The polymer used in these experiments is an electron beam resist, and these structures were formed using a combination of spray-coating and electron beam lithography. The subsequent section considers an extension of this method to define metallic nanostructures by using a lift-off technique. This process is illustrated using the example of aluminum nanowires deposited by electron beam evaporation on single-crystal silicon microcantilevers.

PATTERNING ON MICROMACHINED SILICON CARBIDE PLATES

Figure 1 shows an electron micrograph of plate-mode microresonators developed by Boston Microsystems, Inc. [8].

Figure 1: Electron micrograph of two plate-mode SiC/AlN microresonators developed by Boston Microsystems, Inc. Details of the design and microfabrication of these devices are contained in Ref. [8].

As described in Ref. [8], these structures consist of single-crystal silicon carbide coated with aluminum nitride using molecular-beam epitaxy. Each plate is about 4 μm thick with lateral dimensions of 150 μm x 100 μm. These devices are used as resonant sensors, and patterning nanomaterials on the SiC/AlN plates is of interest for increasing the sensitivity and selectivity of these devices. To this end, the plates were first spray-coated with an electron beam resist (950k PMMA A2, Microchem, Inc.) using EVG101 spray-coater (EVGroup). This resist, which consists of 2 wt% polymethyl methacrylate (PMMA) in anisole, was further diluted with methyl isobutyl ketone (MIBK) so as to reduce the viscosity of the resist and make it amenable for spray-coating. The diluted resist solution contained MIBK and anisole in weight ratio of 3:1. The dispense rate, the nozzle pressure, and the ultrasonic power were maintained at 5 μl/sec, 1000 mbar and 1.4 W, respectively. The spray-coating parameters were optimized for a 4 inch wafer, with the spray-coating being performed from edge-to-edge of the wafer. The number of cycles was varied to achieve the required thickness with minimum surface roughness. The stage was also rotated at 80 rpm, with the direction of rotation (clockwise/anticlockwise) changing at the end of every cycle. The microresonators were then baked at 180 °C for 90 seconds on a hot plate. After 30 cycles of spray-coating, the thickness of resist was measured to be 450 ±120 nm.

The spray-coated SiC/AlN microresonators were then patterned using electron beam lithography. The electron beam plotter used in this work is a modified 30 kV field-emission scanning electron microscope (SU-70, Hitachi) with DEBEN stage controller and Nanometer Pattern Generating System (www.jcnabity.com) to control the beam raster as well as to design the pattern. An electron beam operating at 30 kV, with beam current of 357 pA and line doses of 8 nC/cm and 4 nC/cm, were used to pattern arrays of concentric circles and dots in PMMA, as shown in Figure 2. The patterns were developed at room temperature by immersing the structures for two minutes in a solution that contained MIBK and isopropyl alcohol (IPA) in a ratio of 1:3 by volume. Finally, the specimens were soaked in IPA for 30 seconds and then permitted to dry in air.

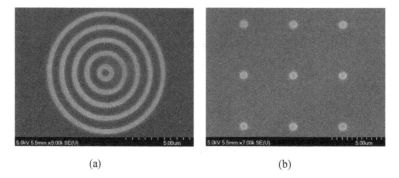

(a)   (b)

Figure 2: Examples of nanostructures patterned using PMMA on the SiC/AlN plate-mode resonators. The darker regions in this image are PMMA and the brighter regions are 500 nm wide trenches exposing the underlying AlN. (a) A set of five concentric circles patterned with an electron dose of 8 nC/cm. The diameters of the innermost and outermost circles are 1 μm and 5 μm, respectively. This pattern was replicated across the surface of the resonator with a center-to-center spacing of 20 μm. (b) Arrays of holes with a diameter of ~500 nm and center-to-center spacing of 5 μm patterned using an electron dose of 4 nC/cm.

PATTERNING ON MICROMACHINED SILICON CANTILEVERS

Metallic nanostructures were patterned on commercial bulk-micromachined Si cantilevers (NSC16/AlBS, Mikromasch) used for scanning probe microscopy. This integration was achieved by first creating polymeric nanostructures using a process similar to the one described earlier. Subsequently, the metal was integrated using a lift-off technique, as described below.

The silicon cantilevers (NSC16/AlBS, Mikromasch) were 230 ± 5 μm long, 40 ± 3 μm wide, and 7 ± 0.5 μm thick. In order to facilitate lift-off, the microcantilever was first spray-coated with a copolymer, MMA-MAA EL11 (Microchem, Inc.). The resist was diluted with MIBK so that the weight ratio of ethyl lactate (EL) to MIBK was 1:1.7, and 5 cycles of spray-coating led to deposition of 390 ± 37 nm of the resist on the microcantilever. The coated cantilever was baked at 150 °C for 90 seconds. This was followed by spray-coating 20 cycles of PMMA A2, which led to deposition of 203 ± 33 nm of the resist. The coated cantilever was again baked at 180 °C for 90 seconds. The bilayer-resist-coated microcantilever was patterned using electron beam lithography with electron doses in the range of 6 nC/cm to 8 nC/cm, and then developed using a solution of MIBK and IPA, as described earlier. Thin films of aluminum were then deposited on the microcantilevers using BJD1800 electron beam evaporator (Temescal) using an operating voltage of 9 kV, emission current of ~400 mA, and a background vacuum of 2×10^{-6} Torr. The deposition rate of Al was 0.2 nm/s. For lift-off, a resist-stripper Remover PG (Microchem Inc.) was heated at 60 °C and the microcantilever was soaked in the heated resist-stripper for 30 minutes. Figure 3 illustrates examples of patterns made by this procedure.

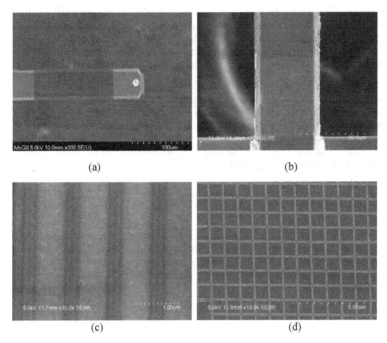

(a)

(b)

(c)

(d)

Figure 3: Electron micrographs of patterns produced on commercial silicon microcantilevers. (a) Selective removal of PMMA from an area measuring 50 μm x 50 μm at the tip and base, exposing the underlying silicon. The darker region in the middle of the cantilever is PMMA-coated. (b) Arrays of nanowires were patterned in an area measuring 60 μm x 40 μm at the base of the microcantilever using an electron dose of 8 nC/cm. This image shows the structure coated with 11 nm of aluminum before lift-off. (c) Images of aluminum nanowires produced after lift-off. Each nanowire is 11 nm thick and 260 ± 3 nm wide, and aligned along the axis of the silicon microcantilever. (d) Checkerboard pattern of aluminum nanowires on a silicon microcantilever produced using an electron dose of 6 nC/cm. Each nanowire is 50 nm thick and 135 ± 2 nm wide, and the centre-to-centre distance between the nanowires is 1 μm.

SUMMARY

The results presented in this paper demonstrate the versatility of an approach that combines spray-coating of electron beam resist, deposition of metallic thin films by evaporation, and electron beam lithography to integrate polymeric and metallic nanomaterials on micromachined silicon and silicon carbide resonators. All the micromachining steps are completed before the synthesis and integration of the nanomaterial. Therefore, the choice of the steps used for deposition and etching, which can require harsh chemical or thermal conditions, are not constrained by compatibility with the nanomaterial. However, some standard processes used for patterning must be modified to account for the low strength of the fragile micromachined components. One example is the use of spray-coating to replace spin coating for dispensing electron beam resist. Our current efforts focus on the use of this method of integration for studying internal friction in nanostructured materials using a microcantilever platform. The details of this platform, and the approach for measuring internal friction, are described in Ref. [9].

In this approach to integrating nanomaterials with MEMS, some of the steps rely upon serial processes. One example is the use of electron beam lithography for patterning. However, the time required for patterning the micromachined structures is comparable to other serial processes, such as wire-bonding, that are routinely used to produce microelectronics and MEMS. Thus, depending on the application, the value added by the nanomaterial can offset the additional time required for the process of integration.

ACKNOWLEDGEMENTS

The authors thank Dr. R. Mlcak of Boston Microsystems, Inc., for providing specimens of the silicon carbide plate resonators, G. Sosale for useful discussions, and the staff of the McGill NanoTools Microfabrication Facility for technical support. Financial support for this work was provided by the Natural Sciences and Engineering Research Council (NSERC) and the Canada Research Chairs program.

REFERENCES

1. M. F. Ashby, P. J. Ferreira and D. L. Schodek, *Nanomaterials, Nanotechnologies and Design*, Butterworth-Heinemann (2009).
2. V. T. Srikar and S. M. Spearing, *Experimental Mechanics*, **43**, 238-247 (2003).
3. M. A. Haque and M. T. A. Saif, *Experimental Mechanics*, **43**, 248-255 (2003).
4. C. Hierold, *Carbon Nanotube Devices*, Wiley (2008).
5. B. E. Alaca, *International Materials Review*, **54**, 245-282 (2009).
6. S. Donthu, N. Alem, Z. Pan, S. Li, G. Shekhawat, V. Dravid, K. Benkstein and S. Semancik, *IEEE Transactions on Nanotechnology* **7** (3), 338-343 (2008).
7. Q. Ye, A. M. Cassell, H. Liu, K. -L. Chao, J. Han and M. Meyyappan, *Nano Letters*, **4**, 1301-1308 (2004).
8. D. Doppalapudi, R. Mlcak, J. LeClair, P. Gwynne, J. Bridgham, S. Purchase, M. Skelton, G. Schultz and H. Tuller in *Microelectromechanical Systems - Materials and Devices III*, Materials Research Society Symposium Proceedings **1222**, Warrendale, PA, 2010, 27-32.
9. G. Sosale, S. Prabhakar, L. Fréchette and S. Vengallatore in *Microelectromechanical Systems - Materials and Devices III*, Materials Research Society Symposium Proceedings **1222**, Warrendale, PA, 2010, 83-88.

Mater. Res. Soc. Symp. Proc. Vol. 1299 © 2011 Materials Research Society
DOI: 10.1557/opl.2011.53

Pt/TiO₂ Growth Templates for Enhanced PZT films and MEMS Devices

Daniel M. Potrepka[1], Glen R. Fox[2], Luz M. Sanchez[1,3], and Ronald G. Polcawich[1]

[1]U.S. Army Research Laboratory, Adelphi, Maryland 20783, U.S.A.

[2]Fox Materials Consulting LLC, Colorado Springs, CO 80908, U.S.A.

[3]Department of Materials Science and Engineering, University of Maryland, College Park, Maryland 20742, U.S.A.

ABSTRACT

The crystallographic texture of lead zirconate titanate (PZT) thin films strongly influences the piezoelectric properties used in MEMS applications. For PZT films poled to saturation, the piezoelectric response is sequentially greater for random, {111}, and {001} texture. Textured growth can be achieved by relying on crystal growth habit and can also be initiated by the use of a seed layer that provides a heteroepitaxial template. Template choice and the process used to form it determine the structural quality and ultimately influence performance and reliability of MEMS PZT devices such as switches, filters, and actuators. This study focuses on how {111}-textured PZT is generated by a combination of crystal habit and templating mechanisms that occur in the PZT/bottom-electrode stack. The sequence begins with {0001}-textured Ti deposited on thermally grown SiO₂ on a Si wafer. The Ti is converted to {100}-textured TiO₂ (rutile) through thermal oxidation. Then {111}-textured Pt can be grown to act as a template for {111}-textured PZT. The Ti and Pt are deposited by DC magnetron sputtering. The TiO₂ and Pt film textures and structure were optimized by variation of sputtering deposition times, temperatures and power levels, and post-deposition anneal conditions. The relationship between Ti, TiO₂, and Pt texture and their impact on PZT growth will be presented.

INTRODUCTION

A wide variety of the physical properties of materials, such as ferroelectricity, ferromagnetism, piezoelectricity, conductivity, and dielectric permittivity depend upon material anisotropy and are therefore strongly affected by chrystallographic texture [1]. With the appropriate choice of thin film texture, device operating efficiency and reliability can be strongly affected. Therefore texture is a critical factor for thin film process control and is fundamental to device reproducibility. This report on deposition and characterization of TiO₂ and Pt thin film electrodes leverages the electrode developments reported for FRAM (ferroelectric random access memory) and applies the technology to PZT-MEMS device fabrication.

Because the Pt electrode crystallographic texture acts as a template for ferroelectric PZT film growth, the remanant polarization properties of the PZT can be degraded by uncontrolled Pt bottom electrode texture [2]. In addition, factors such as bottom or top electrode film density, thickness and grain size and Pt top electrode etch damage and thermal processing influence PZT ferroelectric performance. Academic publications rarely describe Pt texture and density control, but a patented process has been reported for a bottom electrode consisting of {100}-textured TiO₂/ {111}-textured Pt [3,4] which resulted in improved PZT ferroelectric capacitor electrical characteristics for FRAM applications. It is hypothesized that the improved PZT properties result from a reduction in lead loss by suppressing diffusion away from the PZT/Pt interface and templating the PZT crystal growth from the Pt {111} planes, providing enhanced and reproducible {111}-textured PZT.

To ensure a sharp ferroelectric/electrode interface profile and prevent interdiffusion, the bottom electrode must exhibit a high density with a minimum of diffusion pathways (including pores and grain boundaries), and remain stable without degradation of the density and topography. If Pb depletion occurs at the ferroelectric/electrode interface, a low density electrode can allow Pb to diffuse completely through the Pt into the underlying SiO_2 layer. Pb reaction with SiO_2 causes both crystalline and amorphous lead silicate formation, compromising the electronic properties of the substrate and the PZT film [5,6].

Because crystallographic fiber texture is sensitive to composition, density, and surface roughness, texture is an extremely useful parameter for statistical process control of Pt and PZT [7]. In this study, we have used a powder diffractometer to obtain diffraction data for TiO_2 and Pt/TiO_2 thin films on Si substrates. Quantitative analysis of the diffraction data was used to determine the TiO_2 and Pt deposition conditions resulting in the highest level of Pt {111}-texture, which typically corresponds to optimal density and a minimized Pt and Pb diffusion during subsequent PZT processing. The objective of the Pt texture optimization study was to improve orientation, density and the interdependent electrical properties of the subsequently deposited PZT.

EXPERIMENTAL DETAILS

Four inch diameter silicon substrates were obtained with 500 nm $\pm$ 5% thermal oxide from both Addison and Silicon Quest International. Titanium films were sputter deposited on the as-received SiO_2 at a substrate temperature of 40°C using a Unaxis Clusterline 200 chamber configured with a DC rotating-magnet magnetron and a 250 mm diameter, 99.99% pure Ti target. The deposited Ti thin films were converted to TiO_2 with an oxygen anneal in a Bruce Technologies International 6 in. diameter quartz tube furnace. The furnace was heated, with 5 SLM flowing N_2, to the setpoint (650°C – 800°C). Then the wafer boat was loaded and pushed into the tube hot zone over a period of 10 min. At the start of the wafer push step, the N_2 flow was turned off and the O_2 gas flow was ramped up to 10 SLM. After the wafer push, the wafers were held in the hot zone for either 15 or 30 min. The boat was then pulled from the hot zone over a period of 10 min. When the wafers were fully retracted the O_2 flow was turned off, and the N_2 flow was resumed at 5 SLM. The wafers were removed from the boat loader after allowing them to cool for at least 10 min. After the oxidation anneal, Pt was sputter deposited onto the TiO_2 using a second Unaxis Clusterline 200 chamber configured with a DC rotating-magnet magnetron and a 99.99% pure, 250 mm diameter Pt target. All Pt depositions were completed with the same conditions of 0.5 kW, 50 sccm Ar, 30.5 s and a substrate setpoint temperature of 500°C. The resulting Pt films were nominally 100 nm in thickness. To condition the chamber and target, test wafers were deposited prior to sample wafer depositions, thereby ensuring stability of the deposition conditions.

Characterization of the as-deposited metal films included multi-site, 4-point probe measurements of the sheet resistance and single-site X-ray diffraction (XRD) measurements, while characterization of the annealed TiO_2 films included multi-site ellipsometry measurements of thickness and refractive index and single-site XRD measurements. The sheet resistances, R_s, of Ti and Pt films were measured on a Prometrics Versaprobe VP10 and a Four Dimensions, Model 280, Automatic Four-Point Probe Meter, respectively. A J.A. Woollam ellipsometer, model M-2000F, was used for measuring the thickness, and refractive index of the as-received SiO_2 coatings and the annealed TiO_2 films, using a trial thickness of 500 nm (30 nm) for SiO_2 (TiO_2), Si [8] and SiO_2 [9] reference data, and TiO_2 Cauchy layer fitting paramaters $A_n \sim 2.5$ and $B_n \sim 0.05$. A Rigaku Ultima III powder diffractometer with Bragg-Brentano optics, Cu X-ray source, graphite monochromator on the detector arm, and scintillation detector was used to make θ-2θ and rocking-curve measurements on the thin film PZT, Pt, TiO_2 and Ti coated wafers. General θ-2θ measurements were collected using a continuous scan of $2°/min$ and data collection step size of $0.01°$ over a range of 10–$90°$. Using a continuous scan rate of $1°/min$ and step size of $0.02°$, local Ti 200 and 400 and Pt 111 and 222 scans were measured to quantify the position of the maximum θ-2θ peak intensities, immediately followed by a RC θ-scan at a continuous rate of 5-20 sec/count and a step size of 0.02–$0.2°$ for Ti ($1.0°/min$ and step size of $0.01°$ for Pt) at the fixed Bragg angle. A single peak Pearson VII algorithm was used to fit the measured diffraction peaks and obtain quantitative values of RC-FWHM. Because the higher-angle Pt 222 peak provided a lower intensity than the Pt 111 peak, it was used to assure that all quantitative measurements were collected in the linear response range of the diffractometer detector.

DISCUSSION

A basic analysis of the crystallography of Ti, TiO_2, Pt and PZT films was completed by collecting a θ-2θ scan of samples from each deposition condition. Representative spectra are shown in Figures 1, 2a, and 3. Figure 1a shows a θ-2θ XRD spectrum for a Ti thin film deposited at 0.5 kW, 30 sccm Ar, 24 s, and 40°C. The as-deposited Ti film had a nominal thickness of 21 nm and a measured sheet resistance of 89.9 Ohm/sq. Since only two Ti 000*l* reflections, the 0002 and 0004 peaks of the hexagonal-close-packed (HCP) structure (ICDD card #44-1294) are observed in addition to the substrate generated peaks described above, it can be concluded that the Ti film is strongly {0001}-textured with the basal plane of the HCP structure lying in the plane of the substrate. This conclusion was also confirmed by measuring the 0002 rocking-curves of the Ti films and observing RC-FWHM that were consistent with {0001}-texture but varied with deposition conditions. The study presented here focused on finding a combination of Ti deposition and oxidation anneal conditions that resulted in a baseline {100}-textured TiO_2 template for {111}-textured Pt growth and enhanced electrode temperature stability. Therefore, a complete correlation study between Ti deposition conditions and Ti-texture will not be presented. It is conjectured, based on the supporting data collected to date, that the {0001}-texture distribution of the Ti film determines the ultimate TiO_2 {100}-texture distribution that can be achieved via oxidation annealing, but the data presented below indicates that Ti texture is only one factor contributing to the final TiO_2 texture quality.

An example of a diffraction pattern for a TiO_2 film formed by thermal oxidation of the Ti film (annealed at 750°C for 30 min) is shown in Fig. 1b. The TiO_2 film had a measured thickness of 32 nm and refractive index of 2.65. The diffraction pattern exhibits only rutile 200 and 400 peaks, indicative of a {100}-textured TiO_2 film. Variation in anneal temperature (650–800°C) and anneal time (15–30 min) strongly affected the 200 and 400 TiO_2 diffraction peak positions, intensities, and widths but did not cause the appearance of any new diffraction peaks. The quantified parameters of the 200 and 400 TiO_2 peaks reached a plateau at 750–800°C for 30 min anneal times, suggesting completion of the Ti-to-TiO_2 conversion under these conditions.

From these observations, it can be conjectured that the original Ti {0001}-texture is maintained during the oxidation process and that the oxygen merely stuffs the Ti lattice to form the rutile structure in the TiO_2 film with {100}-texture. Such an oxidation mechanism is consistent with the (0001)Ti // (010)TiO_2 epitaxial relationship reported for rutile scale formed on bulk Ti metal [10]. An as-deposited Ti spectrum superimposed on the TiO_2 spectrum shows consistency in how the XRD data is collected for quantitative analysis.

An evaluation of the Pt electrode film sheet resistance did not show any obvious differences related to variations of the TiO_2 seed layer. The sheet resistivity of the Pt was 1.25 Ohms/sq $\pm$ 0.2 Ohms/sq standard deviation. The constancy of the Pt sheet resistance can be expected since all wafers were coated under fixed Pt deposition conditions and because the Pt exhibits isotropic conductivity that is independent of crystallographic texture. Despite the constancy of the Pt sheet resistance, the Pt texture is found to vary significantly and in accordance with the underlying TiO_2 seed layer. An example XRD spectrum of a highly {111}-textured Pt film is shown in Figure 2a, which also shows the superposed spectra for {0001}-textured Ti and {100}-textured TiO_2 seed layers. The expanded y-scale allows one to clearly inspect the background intensity and observe that only the 111 and 222 peaks (the tops of these Gaussian shaped peaks are cut off) occur in the Pt spectrum. In addition to the θ–2θ observations, measurements on the 111 and 222 Pt peak rocking-curves support the conclusion that the Pt film is {111}-textured. Over the range of TiO_2 seed layer thicknesses and processing conditions investigated, the Pt 111 and 222 RC-FWHM varied between 1° and 8°.

In order to prove that the {100}-textured TiO_2 is acting as a seed layer for growth of {111}-textured Pt, a correlation between the texture of the two layers must be established. Such a correlation is shown in Figure 2b, where it can be seen that the Pt 222 RC-FWHM increases linearly with increasing TiO_2 200 RC-FWHM. Additionally, it was observed that for constant thickness TiO_2 and Pt films, an increase in the intensity of the $h00$ TiO_2 diffraction peaks directly corresponds to an increased intensity for the hhh Pt diffraction peaks, which is consistent with grain-to-grain heteroepitaxial behavior.

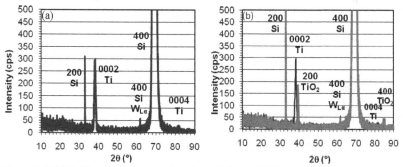

Figure 1. XRD θ-2θ scan of (a) an as-deposited Ti film exhibiting only 000l diffraction peaks and thus confirming the HCP structure Ti {0001}-texture and (b) a TiO_2 film confirming {100}-textured rutile structure and comparison with as deposited {0001}-texture Ti film.

Finally, it was determined that the highly {111}-textured Pt grown on the {100}-TiO_2 seed layer resulted in improved {111}-textured PZT due to templating from the Pt electrode.

Templating was confirmed by comparing the RC-FWHM of the Pt and PZT films. The improved {111}-texture of the PZT was confirmed by comparing the PZT film Lotgering factors of 0.97 and 0.66 for optimized and non-optimized {111}-textured Pt electrodes, respectively. Figure 3 shows the XRD spectrum for a PZT film with Zr/Ti = 0.45/0.55 on the {111}-textured TiO$_2$/Pt electrode superposed onto the spectra of just the Pt and TiO$_2$.

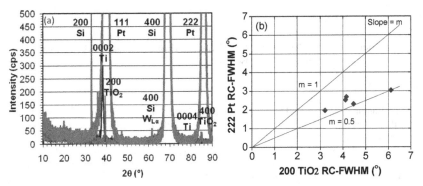

Figure 2. Comparison of XRD θ-2θ spectra (a) for {111}-textured Pt, {100}-textured TiO$_2$ seed layer, and as-deposited {0001}-textured Ti film, and (b) correlation of Pt 222 and TiO$_2$ 100 tilt distribution widths (diamonds), determined from rocking-curve FWHM measurements, and trend reference lines.

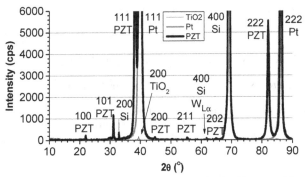

Figure 3. Comparison of XRD θ-2θ spectrum for {111}-textured Pt, {100}-textured TiO$_2$ seed layer and highly {111}-textured Pb(Zr$_{0.45}$Ti$_{0.55}$)O$_3$ prepared with 8% excess Pb.

CONCLUSIONS

Data were presented to confirm that Ti thin films in the range of 10–50 nm thickness can be deposited with a {0001}-textured HCP structure and converted to a {100}-textured TiO$_2$ with the rutile structure by using a thermal oxidation process. The {100}-textured TiO2 can be used as a seed layer to grow a {111}-textured Pt via localized heteroepitaxy. Correlation of the 100 TiO2 and 222 Pt rocking-curve FWHM was used to determine the heteroepitaxy seeding effect. The Pt electrode process change that results in improved {111}-textured Pt has resulted in the {111}-textured PZT growth. The work of Sanchez et al. [11] builds further on this result to show that it has also enabled significant improvements in {100}-textured PZT growth when using a lead titanate seed layer on a Pt electrode surface.

ACKNOWLEDGMENTS

The authors wish to acknowledge Joel Martin and Richard Piekarz (ARL) for their contributions to the processing of wafers used in this study.

REFERENCES

1. M.D. Vaudin, G.R. Fox, and G.R. Kowach, in *Mat. Res. Soc. Symp. Proc., Vol. 721*, edited by P.W. DeHaven, D.P. Field, S.D. Harkness, J.A. Sutliff, L. Tang, T. Thomson, and M.D. Vaudin (Mat. Res. Soc., Warrendale PA, 2002) pp. 17-22.
2. G.R. Fox and S. Sommerfelt, in *Mat. Res. Soc. Symp. Proc., Vol. 721*, edited by P.W. DeHaven, D.P. Field, S.D. Harkness, J.A. Sutliff, L. Tang, T. Thomson, and M.D. Vaudin (Mat. Res. Soc., Warrendale PA, 2002) pp. 145-150.
3. G. Fox and K. Suu, US Patent No. 6 682 772 B1 (27 January 2004).
4. G. Fox, F. Chu, B. Eastep, T. Takamatsu, Y. Horii, and K. Nakamura, US Patent No. 6 887 716 B2 (3 May 2005).
5. H.N. Al-Shareef, D. Dimos, B.A. Tuttle, and M.V. Raymond, J. Mater. Res. **12** (2), 347 (1997).
6. G.R. Fox, S. Trolier-McKinstry, S.B. Krupanidhi, and L.M. Casas, J. Mater. Res. **10** (6), 1508 (1995).
7. F. Chu, G. Fox, T. Davenport, Y. Miyaguchi, and K. Suu, Integr. Ferroelectr. **48**, 161-169 (2002).
8. C.M. Herzinger, B. Johs, W.A. McGahan, and J.A. Woollam, J. Appl. Phys. **83**, 3323 (1998).
9. *Handbook of Optical Constants of Solids Vol. 1*, edited by E.D. Palik (Academic, New York, 1985) p. 759.
10. H.M. Flower and P.R. Swann, Acta Metallurgica **22**, 1339 (1974).
11. L.M. Sanchez, D.M. Potrepka, G.R. Fox, I. Takeuchi, and R.G. Polcawich, in *Mat. Res. Symp. Proc., Fall 2010* (Mat. Res. Soc., Warrendale PA) Manuscript S4.9.

Mater. Res. Soc. Symp. Proc. Vol. 1299 © 2011 Materials Research Society
DOI: 10.1557/opl.2011.54

Contact Resistivity of Laser Annealed SiGe for MEMS Structural Layers Deposited at 210°C

Joumana El-Rifai[1, 2, 3], Ann Witvrouw[1], Ahmed Abdel Aziz[2], Robert Puers[1, 3], Chris Van Hoof[1, 3] and Sherif Sedky[2, 4]

[1]IMEC, Leuven, Belgium.
[2]YJ-STRC, The American University in Cairo, New Cairo, Egypt.
[3]Katholieke Universiteit Leuven, Leuven, Belgium.
[4]Physics Department, The American University in Cairo, New Cairo, Egypt.

ABSTRACT

Lowering the silicon germanium (SiGe) deposition temperature from the current 450°C to below 250°C will enable processing Micro Electro-Mechanical Systems (MEMS) on flexible polymer instead of on rigid silicon substrates or glass carriers. A major disadvantage of such a low temperature deposition is that the films are amorphous, with high hydrogen content and yield poor electrical and mechanical properties. To ensure films suitable for MEMS applications, a post-deposition laser annealing (LA) treatment is used. It is essential that the contact resistance between the SiGe MEMS structural layer and any lower electrode is minimized. In this work we investigate what beneficial effect a LA treatment can have on the contact resistivity of an initially amorphous SiGe MEMS structural layer with a bottom TiN electrode. We report a minimum contact resistivity of 2.14×10^{-3} $\Omega\mathrm{cm}^2$.

INTRODUCTION

Rising interest in processing MEMS on flexible polymer substrates, forces a restriction on the processing temperature, but promises a rise in integration flexibility [1-4]. Using substrates such as benzocyclobutene (BCB), silicone, polyimide (PI) or polyethylene terephthalate (PET) will limit the maximum processing temperature to 250-300°C or lower [3]. A major disadvantage of a temperature reduction beyond 250°C is that the deposition of SiGe at such low temperatures will yield amorphous films with high stress, high stress gradient and high resistivity. All of which are not acceptable for functional and reliable MEMS structural layers. The most common method to produce poly-SiGe or poly-Si films at these low deposition temperatures is a post-deposition laser annealing (LA) treatment [5-7].

The LA technique used in this work employs a local thermal treatment of the material surface using short Excimer laser pulses (~ 24 ns), that maintains the substrate at a low temperature. Previous work demonstrated that the strain gradient of SiGe films deposited at 210°C could become as low as 1×10^{-6} $\mu\mathrm{m}^{-1}$ for a 0.3 μm thick film after LA [6]. For a 1.8 μm thick $Si_{72}Ge_{28}$ layer a strain gradient of -4.3×10^{-6} $\mu\mathrm{m}^{-1}$ was reached after LA [7]. However, in order to achieve these low strain gradients, relatively low laser fluences are recommended and thus only shallow crystallization is achieved. This limits the lowest resistivity that can be achieved to 14 - 80 mΩ·cm [6,7]. In comparison, poly- $Si_{40}Ge_{60}$ deposited using plasma-enhanced chemical vapor deposition (PECVD) at 450°C has a resistivity as low as 1 mΩ·cm [8].

We have previously demonstrated a novel approach for optimizing both the electrical and mechanical properties of the film by selective laser annealing [9]. LA in combination with a patterned Al shielding layer can be used to locally optimize electrical and mechanical properties.

In the contact areas, a resistivity down to 3.47 mΩ·cm can be reached as verified by four-point-probe measurements and in freestanding structures a low strain gradient of -1.6×10^{-7} μm^{-1} is possible [9].

Besides reaching a low resistivity for the SiGe layer, it is also essential that the contact resistance between the MEMS structural layer and any lower electrode is minimized. Work on poly-SiGe layers deposited at 450°C shows that contact resistivity between the poly-SiGe film and the AlCu(0.5%)-TiN interconnect could be as low as $6.2\pm0.4\times10^{-7}$ Ωcm^2[10]. In this work we present results concerning the contact resistivity between laser annealed poly-SiGe and a 100 nm thick bottom TiN electrode.

EXPERIMENT

Using PECVD, SiGe layers were prepared with a fixed Ge content (28%). The film thickness was varied from 0.3 to 1.6 μm and an additional 0.2 μm amorphous Si layer was used to enhance the adhesion to a 1 μm thick SiO$_2$ layer. A thin 100 nm layer of TiN was used as a bottom electrode. The contact between the top Si/SiGe film and the lower TiN layer was achieved through a contact area opening in the SiO$_2$ layer (Figure1a). In a number of samples a soft sputter etch followed by a Ti-TiN (5/10 nm) interlayer deposition was introduced between the SiO$_2$ and Si/SiGe layers (Figure 1b). These steps were performed after the SiO$_2$ etch to improve and clean the contact between Si/SiGe and TiN [10]. Contacts were formed with 1 μm wide trenches and the contact area ranged from a minimum of 1 to a maximum of 236 μm^2(Figure 2a).

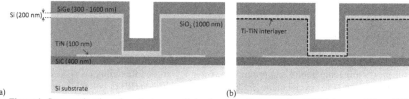

Figure 1- Cross section through contact area outlining the contact between a top Si/SiGe layer to 100 nm TiN electrode (a) without a Ti-TiN interlayer and (b) with a Ti-TiN interlayer.

The local thermal treatment of the a-SiGe structures was conducted with a 248 nm KrF Excimer laser in both air and vacuum. The laser provides a spot size of 23 mm^2 and a pulse duration of 24 ns. Single pulse treatments with maximum energy densities of 280 and 380 mJ/cm^2 in air and vacuum, respectively, were applied. Higher laser energy densities were avoided as they damaged the SiO$_2$ and SiGe layers due to the high hydrogen content in the film which evolves vigorously with high laser energy densities.

The contact resistance was measured after the laser treatment by using a four-terminal cross bridge Kelvin resistor structure (Figure 2b). Measurements involved forcing a current I_2, while measuring the Kelvin potential (V_4-V_1) across bond-pads 4 and 1 (Figure 2b). Mathematically, the contact resistance R_c can be easily calculated using $R_c=(V_4-V_1)/I_2$. Assuming the contact interface is the only contributing factor to the resistance the contact resistivity ρ_c is merely R_cA_c where A_c is the contact area [10, 11].

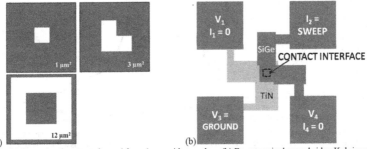

Figure 2- (a) Example of contacts formed from 1 μm wide trenches. (b) Four-terminal cross bridge Kelvin resistor structure used for contact resistance measurements.

The effect of the laser annealing ambient, the laser energy density and the use of a soft sputter etch in combination with a Ti-TiN interlayer on the contact resistivity were all investigated and will be discussed in detail in the following section.

DISCUSSION

As previously investigated, a correlation between the poly-SiGe layer resistivity and the laser annealing condition exists. Increasing the laser energy will lead to a decrease in electrical resistivity and an increase in crystallization depth [9]. Theoretically, a similar effect should be achievable to the contact resistivity. In the next subsections, it will be shown that, as the supplied laser energy density increases up to a maximum value, the contact resistivity is reduced.

We further investigated the role of the annealing ambient. Changing the annealing atmosphere from air to a vacuum of 4 mbar will produce layers with a reduced surface roughness [12] and, as will be discussed in a following subsection, will result in less surface oxidation and radical formation, producing cleaner layers. Finally it is demonstrated that by using a soft sputter etch and a Ti-TiN interlayer it is possible to further reduce the contact resistance.

Laser energy density and contact resistivity

Initial contact resistivity measurements for the as-grown films did not result in any reading, which proved their amorphous state. To establish a good electrical contact between SiGe and TiN, a deep crystallization of the SiGe layer is essential. An increase in supplied laser energy density, implies a deeper crystallization into the SiGe layer and hence a low contact resistivity. Figure 3a illustrates that as the energy density increases up to a maximum energy density of 240 mJ/cm^2, the contact resistivity will decrease. At that energy density the contact resistivity between TiN and a 1.8 μm thick Si/SiGe layer is 3.89×10^{-1} Ωcm^2 for a 48 μm^2 contact and 5.18×10^{-1} Ωcm^2 for a 178 μm^2 contact. Beyond that energy density an increase in contact resistivity is observed up to 280 mJ/cm^2. At higher energy values damages appear in the test structures due to the development of a significantly high tensile stress and high strain gradient caused by the steep thermal gradient across the film thickness.

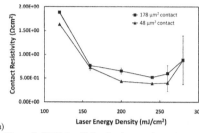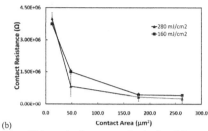

(a)　　　　　　　　　　　　　　　　　　　　　　　　(b)

Figure 3- (a) Plot outlining the decrease in contact resistivity with increasing laser energy density for a 1.8 μm Si/SiGe layer for contact areas of 178 and 48 μm^2. (b) Behavior of contact resistance as a function of contact area for the same sample at laser energy densities of 280 and 160 mJ/cm^2.

Figure 3b shows the behavior of contact resistance as a function of the contact area. As expected, the contact resistance decreases with increasing contact area [11]. Both plots display a larger error for the data obtained with a higher laser energy density of 280 mJ/cm^2 . This is due to the fluctuations and inconsistency in the measurement data, which is a result of the deteriorating test structure because of harsh laser treatment.

These laser annealing experiments are conducted in an air ambient which introduces a large surface roughness due to the oxidation and other radical formation on the surface during the heating process [12]. To reduce this oxidation, for some tests the sample was placed in a vacuum ambient during the annealing process. The effect of such an ambient change is fully discussed in one of the following subsection.

Soft sputter etch and Ti-TiN interlayer

Previous reports on how a soft sputter etch (SSE) and a Ti-TiN (5/10 nm) interlayer can help to reduce the contact resistivity [10], have encouraged us to use the same conditions on our samples. Two samples A and D were prepared with a total thickness of 0.9 μm Si/SiGe. To be able to investigate the effect of a SSE and Ti/TiN interlayer on the contact resistivity of laser annealed samples, sample D was prepared with such conditions, while sample A was not (Table 1). For both samples the contact resistance decreases as a function of the contact area (Figure 4a).

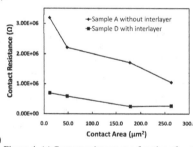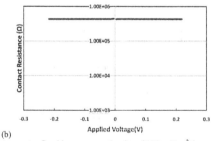

(a)　　　　　　　　　　　　　　　　　　　　　　　　(b)

Figure 4- (a) Contact resistance as a function of contact area at a fixed laser energy density of 120 mJ/cm^2 for samples A and D. (b) Resistance profile for sample D for a 3μm^2 contact area after a 120 mJ/cm^2 laser treatment.

When we compare samples A and D at the same energy density, we find that, for 120 mJ/cm^2, the contact resistance of a 48 μm^2 contact is 2.22×10^6 Ω and 5.91×10^5 Ω for A and D respectively (Figure 4a). Corresponding contact resistivity values are 1.07 Ωcm^2 for sample A and 2.84×10^{-1} Ωcm^2 for sample D (Table 1). The use of the SSE in combination with a Ti-TiN interlayer has also in this work reduced the contact resistivity. This is mainly because this treatment cleans the contact between the TiN bottom electrode and top SiGe layer [10].

Using a SSE and Ti-TiN layer we are able to reduce the contact resistivity down to a minimum of 1.44×10^{-2} Ωcm^2 using the same 120 mJ/cm^2 laser energy density for a 3 μm^2 contact (Table 1). A clear ohmic behavior is observed for this contact (Figure 4b).

Laser annealing ambient

Applying an energy density of 100 mJ/cm^2 to a 1.2 μm thick Si/SiGe layer produces a contact resistivity of 9.05×10^{-2} Ωcm^2 for a 48 μm^2 contact if the laser annealing is conducted in an air ambient (Table 1). Energies higher than 100 mJ/cm^2 will damage the contact and its resistivity will increase. However, changing the annealing environment to a vacuum ambient allows an increase in applied laser energy density up to 160 mJ/cm^2 (Table 1). Under that annealing condition, the contact resistivity of an identical 48 μm^2 contact will decrease to 2.40×10^{-2} Ωcm^2.

The fact that under vacuum conditions a higher laser energy is possible, could be due to various factors. First of all, the energy density that reaches the sample in vacuum could be reduced compared to the applied energy density due to reflection and defocusing effects by the vacuum chamber window. In addition some physical effects might play a role. From the analysis of the SiGe surface by Elastic Recoil Detection (ERD) we observe an increase in surface oxidation and the formation of other radicals in samples that are laser annealed in air (Figure 5). In contrast, samples processed in a vacuum ambient have a limited radical formation. Possibly the cleaner sample surface in vacuum allows for the use of higher energies, but this link is not yet clear. Additionally, previous tests in literature indicated that the laser energy density where the largest grain size occurs is higher for laser annealing in vacuum than in air [12]. In addition, larger grain sizes are seen for treatments in vacuum at this higher energy compared to treatments in air [12].

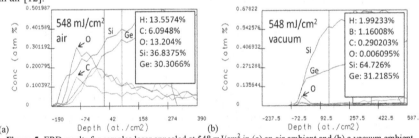

(a) (b)

Figure 5- ERD results for samples laser annealed at 548 mJ/cm^2 in (a) an air ambient and (b) a vacuum ambient.

As previously noted, the contact resistivity for sample D after a treatment in air can reach 1.44×10^{-2} Ωcm^2 for a 3 μm^2 contact. Laser annealing in a vacuum ambient of 4 mbar produces a contact resistivity of 2.14×10^{-3} Ωcm^2 for the same contact area at a higher laser energy density of 200 mJ/cm^2.

Table 1: Overview of the obtained contact resistivity values for samples A-C at similar contact area without the SSE and Ti-TiN interlayer. Sample D illustrates the contact resistivity measured using the SSE and Ti-TiN interlayer.

Wafer ID	Si/SiGe Thickness	20 nm SSE + Ti-TiN	Air Ambient			Vacuum Ambient		
			Laser Energy Density	ρ_c	A_c	Laser Energy Density	ρ_c	A_c
	μm		mJ/cm^2	Ωcm^2	μm^2	mJ/cm^2	Ωcm^2	μm^2
A	0.9	✓	120	1.07	48.0	140	7.68×10^{-1}	48.0
B	1.2		100	9.05×10^{-2}	48.0	160	2.40×10^{-2}	48.0
C	1.8		240	3.89×10^{-1}	48.0	380	1.16×10^{-1}	48.0
D	0.9	*	120	2.84×10^{-1}	48.0			
			120	1.44×10^{-2}	3.0	200	2.14×10^{-3}	3.0

CONCLUSIONS

This work investigates, for the first time, the effect of laser annealing on the contact resistivity between an as-grown amorphous SiGe layer and an underlying TiN electrode. We use cross bridge Kelvin resistance structures to study the contact resistivity. Results show a minimum contact resistivity of 2.14×10^{-3} Ωcm^2 is achievable for a 3 μm^2 contact between TiN and an 0.7 μm thick SiGe layer deposited after a soft sputter etch and a Ti-TiN interfacial layer. For a 4 μm thick boron doped poly-SiGe MEMS layer deposited at 450°C, literature reports a contact resistivity with a minimum value of $6.2\pm0.4\times10^{-7}$ Ωcm^2 for a 2×2 μm^2 contact achieved using a similar SSE and Ti/TiN interlayer. The contact resistivity in the laser annealed samples has room for further improvement by increasing the applied laser energy if we can avoid damage in the surrounding structures. Using selective laser annealing with an Al shield is a possible route for achieving even better TiN-SiGe contacts after laser annealing.

REFERENCES

1. A. Witvrouw, "The road to flexible MEMS integration. In: Passive and Electromechanical Materials and Integration," *Mater. Res. Soc. Symp. Proc.*, vol. 1075, 2008
2. S. Patil et al., "Performance of thin film silicon MEMS on flexible plastic substrates," *Sensors and Actuators A*, vol. 144, pp. 201–206, 2008
3. N. Young et al., "Low Temperature Poly-Si on Flexible Polymer Substrates for Active Matrix Displays and Other Applications," *Mat. Res. Soc. Symp. Proc.*, vol. 769, 2003
4. S. Xiao et al., "A novel fabrication process of MEMS devices on polyimide flexible substrates," *Microelectronic Engineering*, vol. 85 no. 2, pp. 452-457, 2008
5. X. Ma et al., "Laser-Induced Crystallization of a $Si_{1-x}Ge_x$ Microstructure," *J. Mater. Synth. Process.*, vol. 10, no. 1, January 2002
6. S. Sedky et al., "Optimal Conditions for Micromachining $Si_{1-x}Ge_x$ at 210°C", *JMEMS*, vol. 16, no. 3, pp. 581-588, June 2007
7. J. El-Rifai et al., "Laser-induced Crystallization of SiGe MEMS Structural Layers Deposited at Temperatures below 250°C," *Mater. Res. Soc. Symp. Proc.*, vol. 1153, A19.03, 2009
8. M. Gromova et al., "Characterization and strain gradient optimization of PECVD poly-SiGe layers for MEMS applications," *Sensors and Actuators A*, vol. 130-131, pp. 403-410, 2006
9. J. El-Rifai et al., "Selective Laser Annealing for Improved SiGe MEMS Structural Layers at 210°C," *Proc. IEEE MEMS 2010*, pp. 324-327, 2010
10. G. Claes et al., "Improvement of the poly-SiGe electrode contact technology for MEMS," *J. Micromech. Microeng.*, vol. 20, 2010
11. H. Lin et al., "An Evaluation of Test Structures for Measuring the Contact Resistance of 3-D Bonded Interconnects," *ICMTS*, pp. 123-127, March 2008
12. A. T. Voutsas et al., "The Impact of Annealing Ambient on the Performance of Excimer-Laser-Annealed Polysilicon Thin-Film Transistors," *J. Electrochem. Soc.*, vol. 149, no. 9, pp. 3500-3505, 1999

Mater. Res. Soc. Symp. Proc. Vol. 1299 © 2011 Materials Research Society
DOI: 10.1557/opl.2011.252

PZT Thick Films for 100 MHz Ultrasonic Transducers Fabricated Using Chemical Solution Deposition Process

Naoto Kochi[1,2], Takashi Iijima[2], Takashi Nakajima[1] and Soichiro Okamura[1]
[1]Tokyo University of Science, 1-3 Kagurazaka, Shinjuku-ku, Tokyo 162-8601, Japan
[2]National Institute of Advanced Industrial Science and Technology, AIST Tsukuba West, 16-1 Onogawa, Tsukuba, Ibaraki 305-8569, Japan

ABSTRACT

To achieve ultrasonic transducers operating above 100 MHz, square pillar shaped $Pb_{1.1}(Zr_{0.53}Ti_{0.47})O_3$ thick film structures were fabricated using a chemical solution deposition (CSD) process. The fabricated sample showed well-saturated P-E hysteresis curve and butterfly-shaped longitudinal displacement curve. The fabricated samples generated more than 100 MHz ultrasonic waves with a pulser/receiver. Electrical impedance properties of the samples were measured with an impedance analyzer. A number of spurious resonant modes were observed in the frequency range from 40 to 300 MHz. The characteristics of the sample were investigated by finite element method (FEM). The FEM simulations were in good agreement with the experimental results. For free-standing (substrate free) 10-μm-thick PZT film models, the resonant frequency of the thickness vibration mode was estimated to be 160 MHz with the FEM simulations. These results indicate that the substrate affects the behavior of the spurious resonant modes. Therefore, a sample structure was designed using the FEM simulation. The FEM result suggests that the backside of the substrate should be removed to reduce the substrate effects. Consequently, the thickness vibration mode was observed clearly at 160 MHz. This structure is applicable to the micromachined ultrasonic transducers (MUT) operating in the thickness vibration mode above 100 MHz.

INTRODUCTION

Micromachined ultrasonic transducers (MUT) are attractive devices for several applications such as sonars for underwater exploration [1], nondestructive evaluation, and medical imaging systems. There are two types of MUT: capacitive MUT (cMUT) [2, 3] and piezoelectric MUT (pMUT). By comparison with cMUT, pMUT have advantages such as simplicity of fabrication, higher capacitance, and lower impedance [4]. The MUT operating at high frequency have possibilities to enhance the spatial resolution of the medical imaging devices. For example, transducers operating in the range of 20 to 30 MHz have been studied for skin, intravascular, and ophthalmic imaging [5]. In order to observe biological tissue images clearly, ultrasonic waves above 100 MHz are required. However, it is difficult to realize high frequency transducers using the technologies based on bulk ceramic materials because the operating frequency of the transducers generally depend on the geometry of piezoelectric materials such as thickness. Therefore, it is important to develop microfabricated piezoelectric films in the range of several microns. Recently, preparation techniques of the piezoelectric thin films have been well studied for microelectromechanical systems (MEMS) [6, 7]. Among various piezoelectric materials, lead zirconate titanate (PZT) is the most popular material because of its excellent piezoelectric properties, dielectric constant, and thermal stability. It is expected that microfabricated PZT films operating in thickness vibration mode generate high frequency ultrasonic waves compared to typical MUT operating at several MHz in flexural mode [4, 8]

In this study, the PZT thick films for high frequency transducers were fabricated using a chemical solution deposition (CSD) process. The characteristics of the fabricated samples were evaluated in terms of piezoelectric properties and modes of resonant vibrations. Furthermore, finite element method (FEM) simulations were performed to understand the behavior of the ultrasonic transducer and to develop an optimized design of the sample.

EXPERIMENTAL PROCEDURE

Figure 1 shows a schematic illustration and an optical image of the fabricated samples. 200-nm-thick Pt and Ti layers as the bottom electrodes were sputtered onto a 2-inch Al_2O_3 substrate. The prepared $Pb_{1.1}(Zr_{0.53}Ti_{0.47})O_3$ precursor solution was deposited on the substrate by a chemical solution deposition (CSD) process. This method is an effective way to prepare films owing to its advantages of the stability in air, low fabrication temperature, dense microstructure, and good coverage on substrates. Details of the preparation of the precursor solutions are described elsewhere [9]. The sequence of spin coating and pyrolysis at 500°C for 3 min was repeated three times, and then the precursor films were fired at 700°C for 5 min in air. This process was repeated with an automatic coating and firing system composed of a dispenser, a spin coater, and an infrared lamp furnace [10]. The thickness of the PZT films was 10 μm. A 200 nm Pt top electrode was sputtered, and the Pt and the PZT layers were etched by reactive ion etching (RIE) process (Plasmalab 80; Oxford Instruments) using a standard photolithography technique. Consequently, the PZT thick film structures were successfully fabricated. The size of the etched top electrode was varied from 30 X 30 to 1000 X 1000 μm^2. The length and width of the top electrode were equal.

The polarization field (*P-E*) hysteresis curves and the piezoelectric induced longitudinal displacement curves were measured with a ferroelectric test system (FCE-1; Toyo Corporation) connected with a twin beam laser displacement measurement system. This system consists of two individual laser interferometers (MLD-102, MLD-301A; NEOARK). Each of the interferometers measured the displacement of the top electrode and the substrate separately to reduce the effect of the substrate bending [11, 12]. The measurement frequency was 100 Hz, and the amplitude of the bipolar electric fields was 100 kV/cm. A pulser/receiver (Panametrics 5900PR; Panametrics, Inc) was used to generate ultrasonic waves. A single pulse was applied to

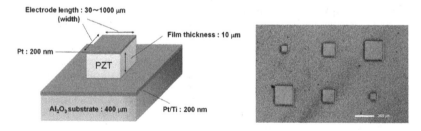

Figure 1. Schematic illustration and optical image of the fabricated PZT thick film structures.

Table I. Material parameters of PZT used in the FEM simulations, where ρ is the density, d is the piezoelectric constant, ε is the electric permittivity, and s is the elastic compliance.

Parameter		Parameter	
$\rho(10^3\text{kg/m}^3)$	7.55	$s_{11}^{E}\,(10^{-12}\text{m}^2/\text{N})$	13.8
$d_{15}(10^{-12}\text{m/V})$	494	$s_{12}^{E}\,(10^{-12}\text{m}^2/\text{N})$	−4.07
$d_{31}(10^{-12}\text{m/V})$	−93.5	$s_{13}^{E}\,(10^{-12}\text{m}^2/\text{N})$	−5.80
$d_{33}(10^{-12}\text{m/V})$	223	$s_{33}^{E}\,(10^{-12}\text{m}^2/\text{N})$	17.1
$\varepsilon_{11}^{T}/\varepsilon_0$	1180	$s_{44}^{E}\,(10^{-12}\text{m}^2/\text{N})$	48.2
$\varepsilon_{33}^{T}/\varepsilon_0$	730	$s_{66}^{E}\,(10^{-12}\text{m}^2/\text{N})$	38.4

the sample, and the sample received the reflected waves from the bottom face of the substrate. An impedance analyzer (E4991A; Agilent Technologies) was used to measure the electrical impedance of the samples. The samples were poled by applying electric field at room temperature just before the measurements.

On the basis of the experiment, two-dimensional FEM simulations were performed with PZFlex as a standard FEM program (Weidlinger Associates Inc) to compare with the characteristics of the experimental results. The models were divided into rectangular shaped meshes. The mesh size was set to 1/20 of smallest wavelength in the simulations. To reduce the computation time, only a half of the model was analyzed with symmetry boundary conditions. Each model was excited by a one cycle input signal at its own central frequency. The variety material parameters of the PZT are shown in Table I [13].

RESULTS AND DISCUSSION

Figure 2 (a) shows P-E hysteresis curve of the sample measured with a ferroelectric test system. The electrode length was 1000 μm. The shape of the hysteresis curve was well saturated with increasing the electric field. The measured remnant polarization and coercive field were $2Pr$ = 37 μC/cm^2 and $2Ec$ = 56 kV/cm, respectively. These values were comparable to those of bulk PZT ceramics. Figure 2 (b) shows bipolar driven longitudinal displacement curve of the sample measured with a twin beam laser displacement system. As can be seen, the sample showed the butterfly-shaped displacement curve, related with piezoelectric response.

Figure 3 shows pulse echo response of the sample. The electrode length was 1000 μm. A single pulse was applied to the sample with a single pulser/receiver. As shown in figure 3 (a), transmitted ultrasonic waves were reflected from the bottom face of the substrate. The interval of the reflected waves was about 80 ns, which corresponds to the propagation time of the ultrasonic waves in the Al$_2$O$_3$ substrate. Figure 3 (b) shows frequency spectrum of the received ultrasonic waves. The time domain of the voltage shown in figure 3 (a) was transformed into the frequency domain by applying the fast Fourier transform. While the highest amplitude wave was observed at 100 MHz, the generated waves were varied from 20 to 200 MHz.

Figure 4 (a) shows the electrical impedance of the samples. The top electrode length was varied from 30 to 1000 μm. As shown in figure 4 (a), the resonant frequencies of lateral vibration modes that change with increasing the top electrode length were observed below 70 MHz. Moreover, a number of spurious resonant modes were also observed between 40 and 300 MHz, and thickness vibration modes were not clear. Figure 4 (b) shows the results of the FEM

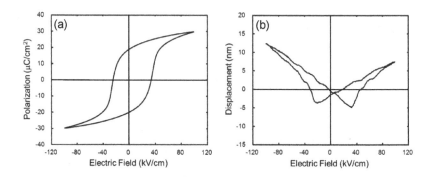

Figure 2. Ferroelectric and piezoelectric responses:
(a) *P-E* hysteresis curve (b) longitudinal displacement curve.

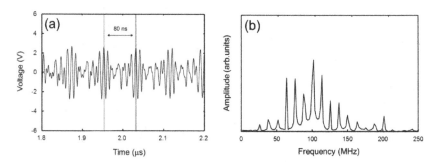

Figure 3. Pulse echo response of the fabricated PZT thick film structures:
(a) time domain response (b) Fourier transform spectrum.

simulations. The analyzed models were the same structure as the fabricated samples shown in figure 1. The boundary condition of the bottom face was fixed while those of the sides were free. The FEM simulations were in good agreement with the experimental results.

Figure 5 shows the FEM simulations of free-standing 10-µm thick PZT structure models which are free from the substrate. In this simulation, all faces of the model were unclamped (free condition). The thickness vibration modes were observed clearly at 160 MHz for all top electrode length, and third order resonant vibration modes and fifth order resonant vibration modes were observed at 500 MHz and 800 MHz, respectively. When the electrode length varied from 30 to 200 µm, the lateral vibration modes and thickness vibration modes overlapped each other around 160 MHz. This means that the electrode that is larger than 500 µm is better to observe the thickness vibration mode. These results suggest that the effects of the substrate cause the spurious resonant mode as shown in figure 4 (a). The spurious modes were induced by the

82

harmonic resonances of the substrate [14]. Therefore, the sample structure that can reduce the substrate effects is required to observe thickness vibration mode clearly.

On the basis of the considerations, a structure of the sample was designed using the FEM simulation. Figure 6 shows the diagram of this structure and results of electrical impedance. The backside of the substrate was removed to reduce the substrate effects. The boundary condition of the bottom of the silicon substrate was fixed while those of the sides were free. The electrode length was determined to be 1000 μm to decrease the resonant frequency of lateral vibration mode below 10 MHz and to avoid the overlapping of the lateral and thickness vibration mode. The spurious resonant modes were reduced and the resonant frequency of the thickness vibration mode was observed clearly at 160 MHz. Therefore, this FEM simulated structure is applicable to the pMUT devices operating in the thickness vibration mode above 100 MHz. This device design should be performed using Si micromachining techniques in the future.

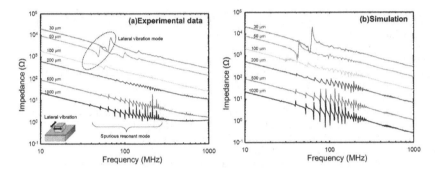

Figure 4. Electrical impedance properties of the fabricated PZT thick film structures: (a) experimental results (b) FEM simulations.

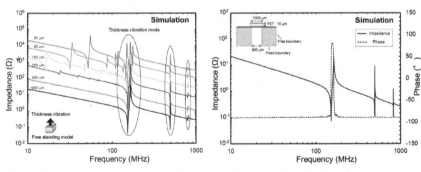

Figure 5. Electrical impedance properties of the free-standing models using the FEM simulations.

Figure 6. Electrical impedance property of the backside substrate removed model using the FEM simulation.

CONCLUSIONS

10-μm-thick PZT film structures for high frequency transducers were successfully fabricated using a CSD process. The fabricated samples generated more than 100 MHz ultrasonic waves. The experimental results of electrical impedance properties were in good agreement with the FEM simulations. A number of spurious resonant modes were observed in the frequency range from 40 to 300 MHz because of the substrate effects. To reduce the substrate effects, a backside substrate removed model was investigated with the FEM simulation. The spurious resonant modes were reduced and the thickness vibration mode was observed clearly at 160 MHz. These results suggest that the 10-μm thick PZT film structure is applicable to the pMUT devices operating in the thickness vibration mode above 100 MHz.

REFERENCES

1. J.J. Bernstein, S.L. Finberg, K. Houston, L.C. Niles, H.D. Chen, L.E. Cross, K.K. Li, and K. Udayakumar, *IEEE Trans. Ultrason. Ferroelectr. Freq. Control.*, **44**(5), 960-969. (1997)
2. B. T. Khuri-Yakub, C. H. Cheng, F. L. Degertekin, S. Ergun, S. Hansen, X-C. Jin, and O. Oralkan, *Jpn. J. Appl. Phys.,* **39**(5B), 2883-2887 (2000)
3. A. Caronti, R. Carotenuto, and M. Pappalardo, *J. Acoust. Soc. Am.,* **113**(1), 279-288 (2003)
4. F. Akasheh, T. Myers, J. D. Fraser, S. Bose, and A. Bandyopadhyay, *Sens. Actuators A*, **111**, 275-287, (2004)
5. P. Maréchal, F. Levassort, J. Holc, L.P. Tran-Huu-Hue, M. Kosec, and M. Lethiecq, *IEEE Trans. Ultrason, Ferroelectr. Freq. Control*, **53**(8), 1524-1533 (2006)
6. S. Trolier-mckinstry, P. Muralt, *J. Electroceram.*, **12**, 7-17 (2004).
7. J. Baborowski, *J.Electroceram.*, **12**, 33-51 (2004)
8. Z. Wang, W. Zhu, J. Miao, H. Zhu, C. Chao, and O.K. Tan, *Sensors and Actuators A*, **130-131**, 485-490, (2006)
9. T. Iijima, N. Sanada, K. Hiyama, H. Tsuboi, M. Okada, *Mater. Res. Soc. Symp.*, **596**, 223-228 (2000)
10. T. Iijima, S. Osone, Y. Shimojo, and H. Nagai, *Int. J. Appl. Ceram. Technol.*, **3**(6), 442-447 (2006).
11. A. L. Kholkin, Ch. Wutchrich, D. V. Taylor, and N. Setter, *Rev. Sci. Instrum.*, **67**(5), 1935-1941 (1996)
12. Y. Kashiwagi, T. Iijima, T. Nakajima, and S. Okamura, *J. Ceram. Soc. Jpn.*, **118**(8), 640-643 (2010)
13. D. A. Berlincourt, C. Cmolik, and H. Jaffe, *Proc. IRE* **48**, 220-229 (1960)
14. J.D.N. Cheeke, Y. Zhang, Z. Wang, M. Lukacs, and M. Sayer, *Proc. IEEE Ultrason. Symp.*, 1125-1128 (1998)

Mater. Res. Soc. Symp. Proc. Vol. 1299 © 2011 Materials Research Society
DOI: 10.1557/opl.2011.253

Reliability and stability of thin-film amorphous silicon MEMS on glass substrates

P. M. Sousa[1], V. Chu[1] and J. P. Conde[1,2]

[1]INESC Microsistemas e Nanotecnologias (INESC MN) and IN-Institute of Nanoscience and Nanotechnology, Rua Alves Redol, 9, 1000-029 Lisboa, Portugal
[2]Department of Chemical and Biological Engineering, Instituto Superior Técnico, 1000-049 Lisbon, Portugal

ABSTRACT

In this work, we present a reliability and stability study of doped hydrogenated amorphous silicon (n^+-a-Si:H) thin-film silicon MEMS resonators. The n^+-a-Si:H structural material was deposited using radio frequency plasma enhanced chemical vapor deposition (RF-PECVD) and processed using surface micromachining at a maximum deposition temperature of 110 °C. n^+-a-Si:H resonant bridges can withstand the industry standard of 10^{11} cycles at high load with no structural damage. Tests performed up to 3×10^{11} cycles showed a negligible level of degradation in Q during the entire cycling period which in addition shows the high stability of the resonator. In measurements both in vacuum and in air a resonance frequency shift which is proportional to the number of cycles is established. This shift is between 0.1 and 0.4%/1×10^{11} cycles depending on the applied V_{DC}. When following the resonance frequency in vacuum during cyclic loading, desorption of air molecules from the resonator surface is responsible for an initial higher resonance frequency shift before the linear dependence is established.

INTRODUCTION

Micro-electro mechanical systems (MEMS) are important components for sensor and actuator applications in which miniaturization is a key aspect as they have the potential of enhancing sensitivity and increasing actuation speed and precision. For MEMS applications, the stability and reliability of device performance are critical and being able to relate these parameters to changes in mechanical properties upon stress is of great importance in the design of these systems. Movable parts in MEMS devices under long-term repeated cycling load are subject to possible failure mechanisms including material fatigue and aging, mechanical fracture, stiction, wear, delamination, residual stress, and environmentally induced failure mechanisms, such as shock, vibration, humidity, particle contamination and electrostatic discharge [1]. In the field of MEMS, industry standard tools and techniques for understanding and quantifying reliability are still limited. However, a general criterium for moving parts is that 10^{11} continuous cycles are required [2]. The frequency stability of a resonator can be given by its quality factor (Q) and how it handles power (applied voltage) [3].

For a brittle material like silicon under applied stress, fracture occurs at the sites of highest stress concentration which usually are processing-related. However, micron-scale fatigue has also been reported for single crystal and polycrystalline silicon [4]. In the case of thin-film hydrogenated amorphous silicon-based MEMS structures the mechanisms that affect device reliability are still an open question and have not been investigated [5].

Hydrogenated amorphous silicon is being investigated as an alternative MEMS structural material [6,7] due to its low temperature processing and the possibility to extend MEMS to applications involving large area substrates such as glass, plastic or flexible metal foils. In this work, we present a reliability and long term stability study of MEMS resonators based on doped hydrogenated amorphous-silicon (n⁺-a-Si:H) thin-films deposited by radio frequency plasma enhanced chemical vapor deposition (RF-PECVD). We investigate the number of cycles to failure and the material aging by monitoring the MEMS properties during long-term cyclic loading.

EXPERIMENTAL

The electrostatic resonators used for the reliability and stability measurements are clamped-clamped bridge structures with 30 μm length and 8 μm width and have an underlying gate electrode. The structures were fabricated on glass substrates at a maximum processing temperature of 110 °C using CMOS-compatible technology. A ~1 μm thick photoresist layer was used as a sacrificial layer. The n⁺-a-Si:H 300 nm thick structural film is deposited at 100 °C by RF-PECVD [8]. The full process flow has been published elsewhere [6]. The micro-resonator structures were electrostatically actuated at room temperature by applying a voltage with DC and AC components [$V_{Gate} = V_{DC} + V_{AC} \sin(2\pi f t)$] between the bridge and the gate electrodes. An AC voltage of 1.26 V was used in all experiments. The resonance frequency of individual bridges was monitored by focusing a laser spot on top of the structure and measuring the deviation of the reflected beam with a photodetector. The photodetector signal was read by a spectrum analyzer [9]. This signal is proportional to the bridge deflection amplitude. The microbridges were electrostatically excited to nonlinear regimes in a frequency interval of up to ±1 MHz centered on the fundamental flexural resonance mode. The resonance peaks were acquired every ~5x10⁸ cycles and fitted to a Lorentzian function from which the resonance frequency, f_{res}, and the quality factor, Q, defined as $Q = f_{res}/\Delta f_{-3dB}$ where Δf_{-3dB} is the resonance peak width 3 dB below its maximum, are extracted. Experiments were conducted at room temperature in vacuum (10⁻⁶ Torr) and at atmospheric pressure.

RESULTS AND DISCUSSION

Figure 1 shows the SEM image of a bridge after being subjected to 10¹¹ loading cycles at 32 V showing no apparent structural physical damage.

Figure 1: SEM image of a n⁺-a-Si:H bridge after being actuated at V_{DC} = 32 V for 10¹¹ cycles. Length: 30 μm, width: 8 μm, thickness: 300 nm, air gap: ~1 μm.

Figure 2 shows the frequency spectrum of a 30 μm long bridge at the fundamental flexural resonance peak (~2 MHz) as a function of applied V_{DC} measured in vacuum. A decrease in resonance frequency and increase in the width and asymmetry of the resonant peak can be observed with increasing V_{DC}. The nonlinear behavior of the resonator, observed for $V_{DC} \geq 8$ V, is a consequence of non-harmonicity resulting from the larger displacements.

Different bridges with the same dimensions were actuated in vacuum with different loads (V_{DC} values) for a fixed V_{AC} of 1.26 V for at least 1×10^{11} continuous loading cycles. As can be seen in figure 3, an increase in the value of the resonance frequency is observed as the number of cycles increases. For $V_{DC} = 16$ V, the resonance frequency increases linearly and after 1×10^{11} cycles, the shift is +0.13%, while for $V_{DC} = 32$ V the shift after the same number of cycles is +0.4%. The full-width at half maximum (FWHM) of the peak, related to Q in a symmetrical peak, remained constant throughout the cycling process and for the different loading values. As an example, the FWHM for $V_{DC} = 8$ V is shown as an inset in fig. 3 as a function of the number of the load cycles.

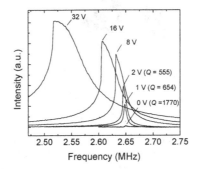

Figure 2: Fundamental flexural resonance peak of a n^+-a-Si:H MEMS bridge as a function of the applied DC voltage. For symmetrical peaks, the value of Q is indicated. The AC voltage was kept at 1.26 V.

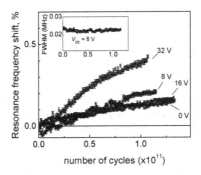

Figure 3: Resonance frequency shift (%) of n^+-a-Si:H MEMS bridges actuated with different DC load values as a function of the number of cycles. The inset shows the peak FWHM as a function of the number of cycles for $V_{DC} = 8$ V. The AC voltage was kept at 1.26 V.

For the higher actuation voltage of $V_{DC} = 32$ V, higher displacements of the bridge may lead to the development of internal stresses and the possibility of damage to the structural material. For this load value, a maximum deflection of 3.3 nm was calculated analytically. In figure 4, the vibration amplitude spectra acquired in vacuum during a continuous cycling experiment is shown. Although the resonance peak has an asymmetrical shape, it does not change significantly throughout the 3.1×10^{11} cycles.

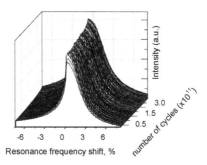

Figure 4: Resonance spectra of a n^+-a-Si:H MEMS bridge as a function of the number of cycles for $V_{DC} = 32$ V. The AC voltage was kept constant at 1.26 V.

The resonance frequency shift relative to the spectra in fig. 4 is plotted versus number of cycles in figure 5 (curve (a)). After 3×10^{11} cycles, the bridge showed a total resonance frequency shift of +0.75%. Most of this shift (+0.44%) occurs on the first 10^{11} cycles. At the end of 3×10^{11} cycles, the change in the resonance frequency shift is ~0.04%/10^{11} cycles as indicated in the plot. Furthermore, the FWHM value remains constant for the full cycling process, as can be seen in the inset of fig. 5 (curve (b)). This result shows that no significant damage or fatigue can be observed in the structure as a result of the cyclic load. The cycle load experiment was also carried out at atmospheric pressure. Since, under these conditions, there are greater variations of ambient humidity and atmospheric pressure, the results show that, besides the expected broadening of the resonance peak due to increased damping, the scatter of values of resonance frequency is much higher (curve (c)). However, the resonance frequency shift observed upon load cycling is similar to that one observed under vacuum. In addition, as can be seen in the inset of fig. 5 (curve (d)), the FWHM value also remains constant throughout the cycling experiment.

The initial, sharper increase in resonance frequency observed in vacuum (fig. 5, curve (a)) could indicate the effect of desorption of air and/or H_2O molecules from the bridge surface, which is not observed in the measurements made at atmospheric pressure. Curve (e) in fig. 5 shows the resonance frequency shift of a bridge that was kept in vacuum for a time equivalent to ~2.2×10^{11} cycles (30 hours) before the load was applied and the resonance frequency measured. The total resonance frequency after 1×10^{11} cycles increased +0.27% instead of the +0.4% observed in curve (a). This suggests the pumping prior to loading reduced the adsorbed species present on the bridge surface. For this bridge, after the initial 1×10^{11} cycles, the resonance frequency shift became linear with the number of cycles (~0.1%/10^{11} cycles).

The effect of the molecular desorption on the resonance frequency during cyclic loading was further studied as shown in figure 6. The same DC load (32 V) was applied to two different bridges. The first bridge was actuated for ~1x10[11] cycles and then kept in vacuum for 8 hours before restarting the actuation and the resonance frequency measured (fig. 6, (a)). The second bridge, after the same number of cycles, was exposed to air for 8 hours after which the bridge was again actuated and the resonance frequency measured in vacuum (fig. 6, (b)). When restarting actuation after 8 hours in vacuum only a small shift is observed. For the bridge exposed to air after each actuation for 1x10[11] cycles a different behavior was observed. After each exposure to air, the resonance frequency shifted to lower values indicating mass loading by the adsorbed molecules. This experiment strongly suggests that the higher slope of the resonance frequency shift observed in the beginning of the load cycling in vacuum can be attributed to the desorption of air molecules from the resonator surface.

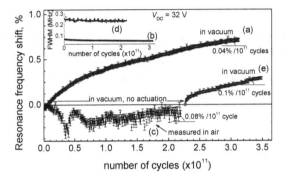

Figure 5: Resonance frequency of a n⁺-a-Si:H MEMS bridge measured in vacuum (curves (a) and (e)) and at atmospheric pressure at V_{DC} = 32 V (curve (c)) as a function of the number of load cycles. Curve (e) corresponds to a bridge actuated after being 30 hours in vacuum without actuation. The inset shows the FWHM as a function of cycles for the sample measured in vacuum (curve (b)) and in air (curve (d)). The AC voltage was kept constant at 1.26 V.

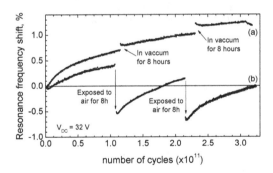

Figure 6: Resonance frequency shift of n⁺-a-Si:H MEMS bridge either periodically exposed to air or held in vacuum for a period without actuation as a function of the number of load cycles.

CONCLUSIONS

Doped hydrogenated amorphous silicon MEMS bridges can withstand the industry standard of 10^{11} cycles at high load (deflection estimated at 3.3 nm) with no observed structural damage. A negligible level of degradation in Q during the entire cycling period shows the high stability of the resonator. In vacuum, the desorption of air molecules from the resonator surface is responsible for the initial faster resonance frequency shift observed. After, this initial stage, the resonance frequency shift becomes proportional to the number of cycles. In air, this linear behavior is observed since the beginning of the load cycle. The linear shift of the resonance frequency is between 0.1 and $0.4\%/1 \times 10^{11}$ cycles depending on the applied V_{DC}.

ACKNOWLEDGMENTS

The authors wish to thank J. Bernardo, F. Silva, and V. Soares for help in cleanroom processing and characterization. Pedro M. Sousa also acknowledges Fundação para a Ciência e a Tecnologia (Portugal) for a post-doctoral grant BPD 44425/2008. This work was supported by the FCT through research projects (PTDC/CTM/72772/2006) and through the Associated Laboratory IN.

REFERENCES

1. W. M. van Spengen, Microelectronics Reliability, **43**, 1049 (2003).
2. Jan G. Korving, Oliver Paul (Editors), *MEMS: a practical guide to design, analysis, and applications*, (William Andrew, inc., 2006, New York) p. 81.
3. W.-M. Zhang, G. Meng and D. Chen, Sensors, **7**, 760 (2007).
4. D. H. Alsem, O. N. Pierron, E. A. Stach, C. L. Muhlstein and R. O. Ritchie, Adv. Eng. Mater., **9**, 15 (2007).
5. J. Gaspar, O. Paul, V. Chu, and J. P. Conde, Mat. Res. Soc. Symp. Proc **1066**, 1066-A15-04 (2008).
6. S. B. Patil, V. Chu and J. P. Conde, J. Micromech. Microeng. **19**, 025018 (2009).
7. J. Gaspar, V. Chu and J. P. Conde, IEEE J. Microelectromechanical Systems **14**, 1082 (2005).
8. P. Alpuim, V. Chu, and J. P. Conde, J. Appl. Phys. **86**, 3812 (1999).
9. J. Gaspar, V. Chu and J. P. Conde, J.Appl. Phys. **93**, 10018 (2003).

Mater. Res. Soc. Symp. Proc. Vol. 1299 © 2011 Materials Research Society
DOI: 10.1557/opl.2011.58

High Yield Polymer MEMS Process for CMOS/MEMS Integration

Prasenjit Ray, V. Seena, Prakash R. Apte, Ramgopal Rao
Centre of Excellence in Nanoelectronics, Department of Electrical Engineering,
Indian Institute of Technology Bombay, Mumbai, India.

ABSTRACT

MEMS community is increasingly using SU-8 as a structural material because it is self-patternable, compliant and needs a low thermal budget. While the exposed layers act as the structural layers, the unexposed SU-8 layers can act as the sacrificial layers, thus making it similar to a surface micromachining process. A sequence of exposed and unexposed SU-8 layers should lead to the development of a SU-8 based MEMS chip integrated with a pre-processed CMOS wafer. A process consisting of optical lithography to obtain SU-8 structures on a CMOS wafer is described in this paper.

INTRODUCTION

The negative photoresist SU8 has been in use for developing high aspect ratio Micro-electromechanical system (MEMS) structures. SU8 structures with aspect ratio up to 25 have been reported[1-5]. Such high aspect ratio structures are used to fabricate devices like cantilevers, accelerometers, pressure sensors, micromirrors etc. Due to the low process temperature of SU8, it is possible to follow the monolithic approach of "CMOS first MEMS last" for realization of System-On-Chip (SOC) applications. One of the processes for fabrication of suspended SU8 structures is the flip-chip release technique [6-7]. Other technique to get hanging SU8 structure on the wafer is to use e-beam lithography with varying exposure energy [8]. One of the more successful methods is the use of Al film between SU8 layer to stop UV light propagation to the sacrificial SU8 layer [9,10]. Some investigators have used Microstreolithography [11], antireflection coating to control the dose[12], and a proton beam [13] to get the complex structures in SU8. A somewhat complex process called the planar self-sacrificial multilayer SU-8 (PSALMS) allows fabrication of hinges and rotating parts in SU-8.[14]. In this paper a process flow to fabricate suspended structures on the substrate using a standard optical lithography is described. A novel concept of including an embedded mask of SU8/Carbon Black has been used to prevent exposure of the sacrificial SU-8 layer. This process has a very high yield and it can be used as a post CMOS process step to fabricate complex SU8 structures for realization of CMOS-MEMS.

EXPERIMENTAL DETAILS

SU8/Carbon black composite is used as an embedded mask to protect the lower layer of SU8 from getting exposed so as to allow fabrication of complex structures of SU8. Sacrificial layers of unexposed SU8 are removed in one single SU-8 developing step at the end of the entire process. In this paper a process for obtaining only two layer structures has been shown but it is possible to fabricate more than two layer SU8 structures using multiple uses of these process steps.

Processing Steps

Process steps for fabricating a suspended structure on silicon wafer are given below,

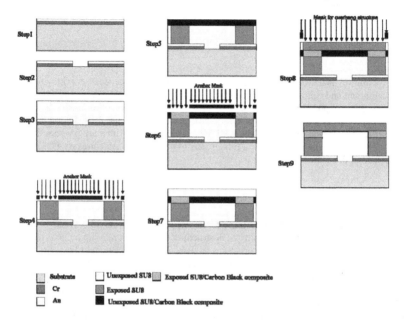

Figure 1. Process Steps for fabricating a suspended structure on Wafer

1. Due to the high transparency between pattern layer and poor contrast between exposed and unexposed layer it is difficult to align subsequent layers of SU8. For this reason high contrast layers like chrome-gold is used for alignment of all subsequent masks.
2. The Cr/Au layer is patterned for alignment marks.
3. The first SU8 layer of SU8-2100 is spun on the wafer with a spin speed of 2200rpm to get a thickness around 150um for providing anchors for suspended structures. Prebake of this layer has been done for 15min at 70°C and for 1hr10min at 90°C.
4. The SU-8 of 150 um thick has been given an adequate exposure of 304mJ/cm^2 to pattern the anchor layer with anchor mask. The post bake of 10min at 70°C and for 25min at 90°C is done without any development..
5. The sample is ready for spinning of SU8/CB embedded mask layer. The SU8/CB(10%) has been found to be optimum for reducing transmittance to ~1%. The SU8/CB

composite is prepared by probe sonication of SU8 2002, Cyclopentanone (SU8 thinner) and Carbon Black for 5min. The SU8/CB is spin coated at a speed of 3000rpm to obtain 2um coating. Prebaking parameter of this layer is 5min at 70°C and for 10min at 90°C.

6. The SU8/CB composite is exposed with a desired area mask (in this case the anchor area) to define the area of SU8/CB embedded mask. The exposure for this layer is done with a larger dose, 608mJ/cm², due to the increased opacity resulting from CB in the SU8/CB composite. All subsequent exposures have to be less than 1/4th of this dose so that the exposed SU8/CB will act as an embedded mask. Post bake of this layer has been done for 5min at 70°C and for 10min at 90°C.

7. Suspended structures of 10 um thickness are obtained by spin coat of SU8 2010 at 3000 rpm and standard prebake at 5min at 70°C and for 10min at 90°C.

8. The suspended structure is defined by exposure through a mask with a dosage of 135mJ/cm² (which is about 1/5th of dose used for SU8/CB composite). This dosage is not sufficient to crosslink the SU8 layer below SU8/CB embedded mask layer.

9. Finally all the structural layers and embedded mask layers have been defined and the wafer is ready for development. The developer removes all unexposed SU8 layer as well as SU8/CB composite layer. The wafer is placed in SU8 developer and agitating by hand for 20min. After this the wafer is double rinsed in isopropyl alcohol (IPA).

The above process steps are sequentially shown in Figure 1.

RESULTS AND DISCUSSION

By introducing Carbon Black (CB) in the SU8 layer transmittance will progressively decrease with increase in CB%. In the process described above, the opaqueness of SU8/CB composite is taken advantage of to prevent the cross linking of the sacrificial SU8 layer below the SU8/CB embedded mask layer.

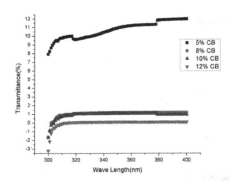

Figure 2. Transmittance of 2um thick film of SU8 on glass substrate at different wavelength for different concentration of Carbon Black in SU8

Fig 2 shows the experimental measurement of transmitted light through 2um thick SU8/CB on glass substrate in the range of 300-400nm. The data is taken for four different weight percentage of Carbon black (5%, 8%, 10%, 12%). The wavelength of 365nm is taken as a reference as it is used in the optical lithography process. The transmittance of the film with 5% CB is 11.275%. If we increase the weight percent to 8%, 10% and 12% then the transmittance will be reduced to 1.21%, 1.08% and 0.113% respectively. This can be compared to the transmittance of 98% for SU8 (without CB). The reduced transmittance of 1% is considered to be enough for making the SU8/CB composite suitable as a masking layer and yet the unexposed SU8/CB composite can be developed by SU8 developer. This embedded SU8/CB mask used as a UV-absorbing layer for the lower layer until a critical dosage is given. So we will get a exposure-time window in which the exposure just penetrates the SU8/CB layer. If the exposure is less than this time then the underlying layer will be unexposed and structure will be hanging. If there is a need for a mechanical connection with lower underlying layer then exposure dosage should be more than this time window. Some of the fabricated suspended structures, namely accelerometer structure (proof mass with 4 beams), torsion structure, meandering spring structure and cantilever beam structure, are shown in Fig3. The large curvature in the structures are due to stress generated by the own weight of the suspended structure. To reduce this stress and curvature we need to reduce the structure size while keeping the thickness same.

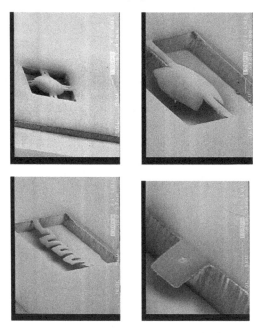

Figure 3. Different fabricated Structures using the SU8/CB embedded mask process

In this process flow only suspended structures of SU8 was fabricated using the SU8/CB embedded mask. By using a modified process flow one can fabricate a SU8/CB comb drive actuator using SU8/CB conducting composite as an actuation layer. For this it is possible to use SU8/CB instead of SU8 in Step 7 and pattern it for suspended comb drive actuator.

CONCLUSION

The novel idea of using a SU8/CB composite as a embedded mask has resulted in a simple process sequence that gives suspended structures in SU8 and SU8/CB composite. In future, using additional process steps, it is possible to integrate piezoresistive SU8/CB or SU8/MWNT layers into a cantilever and accelerometer for realization of piezoresistive sensors. Being a low thermal budget process, all below 90°C, this process is compatible with the CMOS technologies and it will be possible to realize Systems-on-Chip using a "CMOS first MEMS last" approach.

ACKNOWLEDGMENTS

The authors thank the Centre of Excellence in Nanoelectronics for providing the facilities and support. The Centre of Excellence in Nanoelectronics is funded from Ministry of Communication and Information Technology, Govt of India.

REFERENCES

1. Vora K D, Shew B Y, Harvey E C, Hayes J P and Peele A G *J. Micromech. Microeng.* 15 978 (2005).
2. Jiang K, Lancaster M J, Llamas-Garro I and Jin P *J. Micromech. Microeng.* 15, 1522 (2005)
3. Williams J D and Wang W *Microsyst.Technol.* 10, 694 (2004)
4. Zhang J, Chan-Park M B and Conner S R *Lab on a Chip* 4 646 (2004)
5. Chang H-K and Kim Y-K *Sensors Actuators* A 84, 342 (2000)
6. L. Gammelgaard,P. A. Rasmussen, M. Calleja, P. Vettiger, and A. Boisen *Appl. Phy. Lett* 88, 113508(2006)
7. Seena V, Anukool Rajorya, Prita Pant, S Mukherji, V Ramgopal Rao *Solid State Science,* 11, 1606 (2009)
8. Prashant Mali, Aniruddh Sarkar, Rakesh Lal, *Lab on Chip* 6 310(2006)
9. Ceyssens F, Puers R *J Micromech. Microeng.* 16(6) 19(2006)
10. Seidemann V, Rabe J, Feldmann M and Buttgenbach S *Microsyst. Technol.* 8 348(2002)
11. Bertsch A, Lorenz H and Renaud P *Sensors Actuators* A 73 14(1999)
12. Chuang Y-J, Tseng F-G, Cheng J-H and LinW-K *Sensors Actuators* A 103 64(2003)
13. F. H. Tay, J. A. Kan, F. Watt, W. O. Choong *J. Micromech. Microeng.* 11(1) 27 (2000).
14. I G Foulds and M Parameswaran *J. Micromech. Microeng.* 16, 2109 (2006)

Mater. Res. Soc. Symp. Proc. Vol. 1299 © 2011 Materials Research Society
DOI: 10.1557/opl.2011.165

Characterization of Textured PZT Thin Films Prepared by Sol-gel Method onto Stainless Steel Substrates

Xuelian Zhao, Xufang Yu, Shengwen Yu, Jinrong Cheng*
School of Materials Science and Engineering, Shanghai University, Shanghai, China

ABSTRACT

$PbZr_{0.53}Ti_{0.47}O_3$ (PZT) ferroelectric thin films were deposited on $LaNiO_3$ (LNO) buffered stainless steel (SS) substrates by sol-gel method. The effect of LNO buffer layer on the orientation and electric properties of PZT thin films for different thicknesses were studied. X-ray diffraction (XRD) results indicated that PZT thin films on SS substrates exhibit the (100) preferred orientation with the LNO buffer layers. Scanning electron microscope (SEM) images show that PZT thin films were well crystallized with grain size of about 100 nm. PZT thin films deposited on SS maintain the excellent ferroelectric properties with remnant polarization of about 20 $\mu C/cm^2$.

INTRODUCTION

Ferroelectric thin films of PZT in the morphotropic phase boundary (MPB) have attracted much attention for their high values of remnant polarization and dielectric constant as well as low-operating voltage for devices[1]. Most PZT thin films are deposited on platinized Si substrates to be compatible with the semiconductor process. In order to broader the range of application, several groups have already deposited PZT on metal substrates such as Ti[2], Ni[3] and stainless steel (SS)[4]. Advantages of such structures include low cost, the potential for embedded devices and mechanical flexibility of the substrate and so on.

However, the mismatch of lattice and thermal expansion coefficients ,which caused by the interface between the PZT thin films and metal substrates weak the nice property of PZT thin films. Recently, several conductive metal oxide such as $SrRuO_3$ (SRO)[5,6], $YBa_2Cu_3O_7$[7], IrO_2[8] RuO_2[9], $LaNiO_3$ (LNO)[10] and $La_{0.5}Sr_{0.5}CoO_3$ (LSCO)[11] were found not only improve the electrical property but also determine the orientation of PZT thin film. Cheng et al. made investigation on $PbZr_{0.52}Ti_{0.48}O_3$ thin films on a $PbTiO_3$-coated stainless steel substrate and concluded that the intermediate thin PT layer plays an important role in lowering the annealing temperature required for crystallization[12].

PZT thin films with preferred orientation have been prepared by radio frequency (RF) sputtering, pulsed laser deposition (PLD)[13], metal organic chemical vapor deposition (MOCVD)[14] and sol-gel method[15]on single crystal substrates. Compared with other methods, the sol-gel method has several advantages include its simple process, easy control of the composition, low cost of the raw material and so on.

In this work, we aimed to prepare textured $PbZr_{0.53}Ti_{0.47}O_3$ thin films by the sol-gel method onto SS (304) substrates. LNO was chosen as the buffer layer between PZT and SS due to their lattice similarity. The microstructure and electric properties of the films were investigated.

EXPERIMENTAL DETAILS

The LNO sol were prepared by using lanthanum nitrate and nickel nitrate as raw materials, which were mixed in a molar ratio of 1:1 dissolved in the mixed solvents of acetic acid and de-ionized water to obtain LNO sol of 0.1 mol/L. The 7 wt% polyvinyl alcohol (PVA) was added to the system to stabilize the solution. The 0.5 mol/L PZT precursor was prepared by dissolving leadacetate, zirconium isopropoxide and titanium tetrobutoxide into 2-methoxyethanol solvent with 20 mol% excess of lead. The LNO layer with thickness of 40, 60, and 80 nm was deposited on SS substrates with a rapid thermal annealing at 750 °C for 10 min under the oxygen atmosphere. PZT thin films with different thicknesses were deposited on the top of LNO, and then annealed at 650 °C for 30 s in air.

The gold electrode with diameter of 0.4 mm were sputtered on the top of PZT thin films to fabricate the metal-insulator-metal (MIM) capacitor for dielectric measurements. The crystal structure of PZT thin films were obtained by X-ray diffraction (XRD). The morphology of surface was observed by the scanning electron microscopy (SEM). The thickness of the films was determined by an Alpha-step-500. Radiant Technology Ferroelectric tester (Precision Premier II) was used to measure the typical ferroelectric hysteresis loops and the J-V curve using the step delay and soak time of 100 ms.

DISCUSSION

Figure 1(a) exhibits XRD patterns of LNO thin films for different thicknesses. It can be seen that LNO thin layers annealed at 750 °C for 10 min form the perovskite structure with the (110) preferred orientation. No significant variation of the crystalline structure can be observed for LNO films of different thicknesses and the intensity of (110) diffraction peak increases with the increase of the thickness of LNO thin films. In following work, the LNO thin film with thickness of 60 nm was chosen as the buffer layer for PZT thin films. Figure 1(b) exhibits XRD patterns of PZT thin films of 400, 600, and 800 nm deposited directly on SS substrates. It can be seen that PZT film films have the perovskite structure with the random orientation. Using the formula $\alpha_{110} = I(110)/ \text{Sum}[I(\text{all PZT reflections})]$, where I is the XRD peak intensity, the degree of orientation of the (110) plane was calculated to be 48%. However, different results were observed when an LNO layer was deposited. Figure 1(c) presents XRD patterns of PZT thin films with the LNO buffer layer of 60 nm. The (100) textured PZT thin films were obtained on SS substrates upon the LNO buffer layers. The enhanced (100) peak is the consequence of the diminish of the mismatch between the PZT and SS. The interplanar distant of the (100) of the PZT (about 0.403 nm) is closed to the LNO lattice constant (0.384 nm) ,which result in the highly (100) orientation of PZT thin film on LNO buffer layers. The (100) and (200) diffraction peaks of PZT thin films become sharper and stronger with the increase the thickness of PZT thin films, suggesting that the higher PZT thickness facilitates the growth of the (100) oriented PZT films, which lies in the effect of mismatch between LNO and PZT gets weaker.

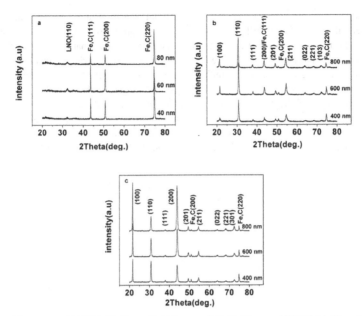

Figure 1. XRD patterns of (a) LNO thin films with thicknesses of 40, 60, 80 nm, and (b) PZT thin films of different thicknesses without LNO, and (c) PZT thin films of different thicknesses with LNO buffer layer of 60 nm

Figure 2. Surface SEM images of (a) LNO thin film with thickness of 60 nm, and (b) PZT thin films with LNO, and (c) PZT thin films without LNO buffer layer of 60 nm

Figure 2 shows the surface SEM images of LNO thin films with thickness of 60 nm and PZT thin films with and without the LNO buffer layer. The microstructure of LNO thin films is uniform with a average grain size of about 60 nm. All the PZT thin films exhibit dense surface and the average grain size range from 50 to 150 nm. PZT thin films without LNO buffer layer have a larger grain size than that of PZT thin films with LNO.

Figure 3 shows the leakage current density of Pt/ PZT/SS and Pt/LNO/ PZT/SS as a function of applied field. According to this figure, the PZT films with LNO layer have a lower leakage current density of 1.07×10^{-7} A/cm^2 under the field of 25 kV/cm than the one without buffer layer which is 5.0×10^{-6} A/cm^2. The decreasing leakage current density result from LNO buffer layer can prevent the reaction of PZT and substrate, which produces conductive oxide phases responsible for the high leakage of the PZT films. The figure indicate that the LNO buffer layer could improve the interface quality. The leakage current in a ferroelectric film can be explained by several conduction mechanisms including Schottky emission, Poole-Frenkel emission and space-charge-limited conduction (SCLC) etc al [16]. The figure 3 shows that J was linearly dependent on electrical field (E) at lower part of the electric fields (E < 10 kV/cm), which indicating that ohmic conduction is the predominant factor in determining the current. When increasing the E, the leakage current density depends linearly on the square root of the applied field ($E^{1/2}$), which is interpreted as the Schottky emission for the conduction mechanism.

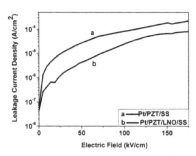

Figure3. Leakage current density of PZT thin films with thickness of 600 nm

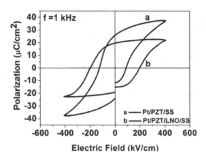

Figure 4. Ferroelectric hysteresis loops of Pt/PZT/SS and Pt/PZT/LNO/SS at the frequency of 1 kHz

The ferroelectric hysteresis loops of Pt/PZT/SS and Pt/PZT/LNO/SS are shown in figure 4. At the frequency of 1 kHz the P–E loop of the PZT with LNO layer is more saturated, which is in accordance with figure 3, the more saturated hysteresis loops of the film may due to its better crystallization, which can be confirmed by the images of figure 2. The remnant polarization (Pr) decreased and coercive field (Ec) increased for the PZT deposited on LNO buffer. The value of Pr with LNO buffer is 19.79 μC/cm^2. LNO buffer layer is a conductor with no Pr , which is the main reason for the decrease of Pr and increase of Ec. The hysteresis loops of the PZT thin films with LNO buffer layers is more symmetric, which can be seen from the separation of the loop at the E = 0 kV/cm in figure 4. The result indicating that LNO buffer layer could improve the interface between PZT thin films and SS substrate.

CONCLUSIONS

Ferroelectric PZT thin films were fabricated on LNO/SS by sol–gel method. The electrical properties of the films with a LNO buffer layer were improved, the introduced LNO layers play an important role in the combination of PZT thin films and SS substrate. Highly (110) oriented LNO thin films on SS substrate were obtained. And with the LNO buffer layer, PZT thin films have a highly (100) and (200) preferred orientation, which exhibit a more saturated ferroelectric hysteresis loop with Pr of 19.79 $\mu C/cm^2$ and lower leakage current density of 1.0×10^{-7} A/cm^2 with the electric field of 25 kV/cm.

REFERENCES

[1] B. Willeng, M. Kohli, K. Brooks, P. Muralt and N. Setter, *J.Ferroelectrics* **201**, 147(1997).
[2] D. J. You, W. W. Jung, S. K. Choi, and Yasuo Cho, *Appl. Phys. Lett.* **84**, 3346 (2004).
[3] Z. L. Hea, Y. G. Wang and K. Bia, *Solid State Communications* **150**, 1837-1839 (2010).
[4] T. Fujii, Y. Hishinuma, T. H. Mita and T. Naono, *Sensors and Actuators* **163** , 220-225(2010).
[5] Y. K. Wang, T. Y. Tseng and P. Lin, *Appl. Phys. Lett.***80**, 3790–3792 (2002).
[6] T. Morimoto, O. Hidaka, K. Yamakawa, O. Arisumi, H. Kanaya, T. Iwamoto, Y. Kumura, I. Kunishima and S. Tanaka, *J. Appl. Phys.* **39**, 2110–2113(2000).
[7] J. Lee, L. Johnson, A. Safari, R. Ramesh, T. Sands, H. Gilchrist and V. G. Keramidas, *J. Appl. Phys. Lett.***63**, 27–29(1993).
[8] T. Nakamura, Y. Nakao, A. Kamisawa and H. Yakasu, *J. Appl. Phys.* **33**, 5207–5210(1994).
[9] S.D. Bernstein, T.Y. Wong, Y. Kisler and R. W. Tustison, *J. Mater. Res.* **8**, 12–13(1993).
[10] B. G. Chae, Y. S. Yang, S. H. Lee, M. S. Jang, S. J. Lee, S. H. Kim, W. S. Beak and S. C. Kwon, *Thin Solid Films* **410**, 107–113(2002).
[11] R. Ramesh, H. Gilchrist, T. Sands, V. G. Keramidas, R. Haakenaasen and D. K. Fork, *J. Appl. Phys. Lett.* **63**,3592–3594(1993).
[12] J. R. Cheng, W.Y. Zhu, N. Li and L. E. Cross. *J. Appl. Phys. Lett.* **81**, 25(2002).
[13] M. D. Nguyen, H. Nazeer, K. Karakaya, S. V. Pham, R. Steenwelle, M. Dekkers, L. Abelmann, D. H. A. Blank and G .Rijnders, *J. Micromech. Microeng.* **20**, 085022(2010).
[14] K. Nishida, M. Osada, S. Yokoyama and T. Kamo, *Key Engineering Materials* **135-138**, 421-422 (2010).
[15] Y. R. Lin, C. Andrews and H. A. Sodano, *Proc. SPIE*, **7644**, 76440C (2010).
[16] G. W. Pabst, L. W. Martin, Y. H. Chu and R. Ramesh, *J. Appl. Phys. Lett.* **90**, 072902(2007).

Micro- and Nanosensors

Mater. Res. Soc. Symp. Proc. Vol. 1299 © 2011 Materials Research Society
DOI: 10.1557/opl.2011.67

A Picowatt Energy Harvester

Joe Evans[1], Johannes Smits[2], Carl Montross[1], and Gerald Salazar[1]
[1]Radiant Technologies, Inc., 2835D Pan American Fwy NE, Albuquerque, NM   87107 USA
[2]Scaldix B V Stationsstraat 13 4331 JA Middelburg, The Netherlands

ABSTRACT

The authors describe an energy harvester circuit fabricated with integrated thin ferroelectric film capacitors on a silicon substrate. The harvesting mechanism is a folded double-beam cantilever with proof masses at both end points. Interdigitated electrode capacitors are located at the three points on the folded cantilever that are expected to experience maximum bending moment and should produce up to 5V as a function of external vibration. The die has the dimensions of 1.6mm on a side and is designed to be mounted in a TO-18 package transistor-style package. Due to its small size, the self-contained piezoelectric MEMs device should produce 50 picowatts in a 1g vibration environment while occupying little space.

THEORY

The authors are fabricating an energy harvesting device with piezoelectric bimorph cantilevers to convert ambient vibration into electrical energy. The device consists of a single double-beam cantilever with a cross-bar tip mass connecting the ends of the two cantilevers, each of which has an interdigitated electrode piezoelectric capacitor at the point of maximum bending. A second single-beam bimorph is placed in the gap between the two beams of the double-beam cantilever. The single-beam cantilever will have a different resonant frequency than the double-beam cantilever and will help capture more energy from the ambient vibration without increasing the overall size of the device. The theory describing energy capture by a piezoelectric bimorph follows.

We calculate the voltage developing at the electrodes of a piezoelectric bimorph vibrating with amplitude a_0 at the clamping point in a direction perpendicular to the length. One of the boundary conditions for the solution of the differential equation is that the clamping point undergoes a sinusoidal motion with an amplitude a_0. All other boundary conditions are those of free bimorphs [1].

The solution of the Euler Bernoulli equation gives the local deflection in the x direction along the length of the cantilever as:

$$z(x) = \left(\frac{a_0}{2(1+cch)}\right)[(1 + cch - ssh)\cos kx + (sch + csh)\sin kx + (1 + cch + ssh)\cosh kx - (sch + csh)\sinh kx]$$

(1)

We use here the conventions $c=cos\,kL$, $s=sin\,kL$, $ch=cosh\,kL$, $sh=sinh\,kL$ as described in [1]. The shape of the vibrated bimorph near resonance is shown in Figure 1.

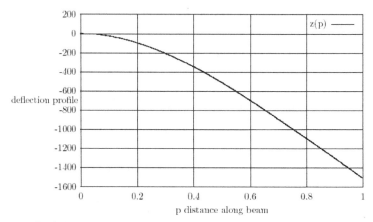

Figure 1. *The deflection profile of the cantilever near resonance.*

Note that the cantilever has a high radius of curvature near the clamping point and is nearly straight at the free end. The piezoelectric capacitor used as the scavenging engine will gain capacitance linearly with its length but will receive minimal piezoelectric distortion near the free end of the cantilever. Consequently, the authors elected to restrict the length of the piezoelectric capacitors to the region of maximum bending so that each cantilever will operate a higher energy to capacitance ratio.

To find the charge induced between the electrodes we use the piezoelectric constitutive equations in the form given by Berlincourt, Curran and Jaffe [2].

$$T = c^E S - eS$$
$$D = eS + \varepsilon^S E \tag{2}$$

S is the elastic deformation, or strain, T the elastic stress, D is the dielectric displacement, and E the electric field. The elastic stiffness modulus at constant electric field is c^E, the piezoelectric constant is e, and the dielectric constant at constant strain is ε^S. The charge Q under short circuit conditions, ($E=0$) in the electrodes becomes:

$$Q = \int_0^L w D_3 dx = \frac{w e_{31} h a_0 k(s-sh)}{1+cch} \tag{3}$$

We use the symbol w to represent the width of the bimorph, h to represent the thickness of a single piezoelectric element of the bimorph, and k to represent the wave number of the driving vibration.

Voltage and charge are related by $Q=CV$, where C is the capacitance. For the capacitance we have [3]:

$$C = \frac{2 \varepsilon_{33}^T L w}{h}\left(1 - k_{31}^2/4\right) \tag{4}$$

where L is the length of the bimorph. Combining equation (3) and (4) we arrive at

$$V = \frac{e_{31} h^2 a_0 k(s-sh)}{2 e_{22}^T L \left(1 - \frac{k_{31}^2}{4}\right)(1+cch)} \tag{5}$$

The total electrical energy U_e inside the bimorph is now $U_e = \{ QV/2 \}$

$$U_e = \frac{1}{4} \frac{w e_{31}^2 h^3 a_0^2 k^2}{e_{22}^T L \left(1 - \frac{k_{31}^2}{4}\right)} \frac{(s-sh)^2}{(1+cch)^2} \tag{6}$$

Mechanical energy is introduced into the bimorph by a vibrating clamping fixture. In each place we can cut the beam into two pieces, and represent the forces on the cut planes which the two pieces exert on each other. The force F_s in the planes is equal to $F_s = -EIz'''$, so the force at the clamping point: $-EIz'''(0)$. The mechanical energy put into the beam is $\{F_s(0) a_0/2\}$. For the mechanical energy U_m we find:

$$U_m = \frac{w h^3}{12 s_{11}} \left(\frac{a_0^2 k^3 (sch+csh)}{2(1+cch)}\right) \tag{7}$$

With the electrical and mechanical energies, we can calculate the energy conversion efficiency η:

$$\eta_e = \frac{U_e}{U_m} = \frac{k_{31}^2}{\left(1 - \frac{k_{31}^2}{4}\right)} \frac{6(s-sh)^2}{kL(1+cch)(sch+csh)} \tag{8}$$

For a bender operating in a $d_{(33)}$ mode the factor $k_{(31)}$ should be replaced by $k_{(33)}$.

The internal energy, mechanical and electrical, must reach steady state before the electrical part can be removed. The steady state total internal energy is equal to the total amount of energy that has been put in $U_i = U_e + U_m$. The electrical energy can be taken out as output energy U_o. We define power conversion efficiency as

$$\eta_p = \frac{U_0}{U_i} = \frac{U_e}{U_e + U_m} = \frac{\eta}{\eta+1} \tag{9}$$

Because the energy conversion efficiency reaches very high values near resonance, the power conversion efficiency approaches the value of unity at resonance and it becomes advantageous to design the energy harvester with more than one resonance frequency, as in the combination of a double-beam and single-beam cantilever pair.

DEVICE FABRICATION

The fabrication process for the energy scavenger combines a well-established thin PZT film capacitor process with known micro-electromechanical machine (MEMs) process steps. Both process types use integrated circuit techniques which, when combined, allow the definition of structures geometrically on CAD system software and the transference of the design to a substrate surface with micron tolerances. The conversion of a piezoelectric MEMs concept to real machine simplifies in this situation to integration of the plan forms of PZT capacitors with those of the micro machinery. The concept of a geometric approach to micro-scale devices was introduced to the VLSI integrated circuit design discipline (Very Large Scale Integration) in 1980 by Mead and Conway [4], an advance which resulted in an explosion of commercial IC circuits, computers, and automated circuit layout tools. The authors hope that this same philosophy will benefit piezoelectric MEMs technology in the same way.

Capacitor Process:

The capacitor process in use for the fabrication of the piezoelectric elements uses capacitors having thin 4% niobium-doped 20/80 PZT (PNZT) films and platinum electrodes. The interdigitated electrodes of Figure 1 require only one level of platinum in the process flow. The electrodes for this device were executed on top of the thin PNZT.

The process uses a commercial lift-off step to pattern the platinum electrodes. The top platinum is 1000Å in thickness, making it possible to achieve one micron features and gaps with the lift-off process. Despite the presence of platinum electrodes with the PNZT, fatigue is not a factor because the polarization is never switched after the first poling. Imprint may over time affect the energy generation efficiency. The niobium doping of the 20/80 PZT significantly reduces the imprint rate of the domains for temperatures below the industrial standard of 85°C.

Standard industrial procedures such as etching, sawing, soldering, and packaging after the capacitors have been completed have an impact on film performance. Passivation is critical to protect capacitor performance during the MEMs etch processes. The passivation for these capacitors places a stack of titanium dioxide and silicon dioxide on their top surfaces, essentially adding 2,700Å of glass atop the piezoelectric elements.

MEMs Process:

The energy scavenger has three thicknesses in its mechanical design. The first is the substrate thickness itself, an etch depth of zero from the backside of the substrate. The substrate is necessary to 1) provide structural strength to the micromachine, 2) provide a foundation to mount the device in packaging, and 3) create inertial masses. The second thickness is that of the cantilevers. The final thickness is that of the full substrate so the device can be self-diced from the substrate after the last etch.

Two deep RIE steps and a buried oxide layer in the substrate (SOI) are used to create the MEMs structure. The depth of the buried oxide determines the thickness of the cantilevers. After completion of the capacitors, a trench pattern is placed on the top surface of the substrate in photoresist. A DRIE etch is executed to the buried oxide. This pattern defines the geometry of the cantilevers as well as the dicing lanes. The substrate is flipped over, a backside pattern is placed on the backside of the substrate, and a second DRIE etch is executed to the buried oxide. The backside pattern is not the same as the trench pattern from the front side. Where the backside and trench patterns align, the substrate will be etched through to the surface to form the dicing lanes and the lateral dimensions of the cantilevers. Where the backside and trench patterns *do not* align, the substrate is removed from underneath a cantilever, a membrane, or bridge structure. Once the two DRIE process steps have been completed a short plasma etch or wet buffered oxide etch will remove the exposed buried oxide layer to free the dice and the cantilever structures.

PHYSICAL IMPLEMENTATION

A complete die with dicing lanes appears in Figure 2 defined by the dashed box.

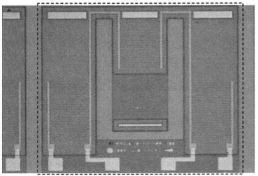

Figure 2. *Photomicrograph of energy scavenging device in fabrication.*

A portion of a second die appears to the left of the complete die. The "U" shaped structure in middle of the die will extend through the wafer to define the cantilevers. The two outside arms, containing the interdigitated electrodes (IDEs), form the double-beam cantilever and will bend together. A crossbar joins the two primary cantilevers at the top of the figure. A single-beam cantilever extends from that crossbar back towards the base of the die exactly in the center of the die. This cantilever has its own IDE set at the point of maximum bending. The substrate underneath the three cantilevers will be removed by the two part etch process, making the cantilevers thin and flexible. The etch process will also release each die from the substrate. Blocks of the original substrate will be left at the ends of the primary and secondary cantilevers to act as inertial masses to increase the amplitude of cantilever tip motion. The gaps between the fingers of the IDE electrodes are 10 microns. The IDEs rest on the top surface of 0.8 micron-thick 4/20/80 PNZT film. They will require 190 volts to pole each capacitor. Connected in series, the three IDE capacitors should generate as much as 5 volts at maximum bending. Were the same capacitors to be fabricated as parallel plate capacitors, they might generate at most 100mV at maximum bending, not nearly enough to power control circuitry to collect and control energy coming from the device. The hysteresis loop for the device in Figure 1 is shown in Figure 3 below.

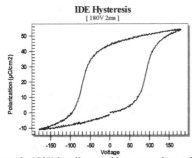

Figure 3. *180V 2 millisecond hysteresis loop of the interdigitated electrodes of Figure 1.*

The die is 1.6mm on a side and will fit inside a four-pin TO-18 package. The TO-18 package is the traditional metal transistor package in use since the 1960s. Figure 3 provides a visual interpretation of the interconnected bending moments of the primary and secondary cantilevers. The die will have to be mounted on a pedestal to raise the inertial masses above the header surface.

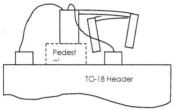

Figure 4. *Side view of the energy scavenger mounted on a pedestal on the TO-18 header. The increased height of the pedestal is necessary to allow full motion of the substrate-thick end-point masses.*

PREDICTION OF PERFORMANCE

The double-beam cantilevers of the energy scavenger of Figure 2 have the dimensions of 1060μ long, 465μ wide, and 40μ thick. The single-beam cantilever has the same thickness and is 880μ long and 470μ wide. The cross bar connecting the two primary cantilevers is 550μ thick, 140μ wide, and 1600μ long and acts as an inertial mass. The inertial mass for the single-beam cantilever is 550μ thick, 470μ wide, and 130μ long. Given that the IDEs have 10μ gaps and 5μ fingers on 0.8μ-thick 4/20/80 PNZT with a d_{33} value of 60pm/v, we predict that resonance will occur at approximately 50 Hz. A 1g sine wave stimulation at resonance should generate 50 picowatts of power and 200mV across the IDE electrodes. The voltage generated by the scavenger will increase linearly with the amplitude of the stimulation (Eq (5)). The power generated will increase with the square of the driving amplitude (Eq(6)).

CONCLUSION

The authors have designed and are fabricating a picowatt energy scavenger using piezoelectric MEMs process technology. The scavenger should approach 100% energy conversion efficiency near mechanical resonance and should be able to generate 50 picowatts of power at 50 Hz/1g.

REFERENCES

1. J. G. Smits and A. Ballato, J Micro ElectroMechanical Systems **3**, 105-112 (1994)
2. A. Berlincourt, D. R. Curran, and H. Jaffe, *Physical Acoustics* D **1** (A), Edited by W. P. Mason (Academic Press, New York, 1964) 188
3. J. G. Smits, S. I. Dalke, and T. K Cooney, Sensors and Actuators **28**, 41-61 (1991)
4. C. Mead and L. Conway, *Introduction to VLSI Systems* (Addison-Wesley, California, 1980)

Mater. Res. Soc. Symp. Proc. Vol. 1299 © 2011 Materials Research Society
DOI: 10.1557/opl.2011.255

Mechanical and Material Characterization of Bilayer Microcantilever-based IR detectors

I-Kuan Lin[1,2], Ping Du[3], Yanhang Zhang[3] and Xin Zhang[3]
[1] Global Science & Technology, Greenbelt, MD 20770, USA
[2] NASA Goddard Space Flight Center, Greenbelt, MD 20771, USA
[3] Department of Mechanical Engineering, Boston University, Boston, MA 02215, USA

ABSTRACT

Infrared radiation (IR) detection and imaging are of great importance to a variety of military and civilian applications. Microcantilever-based IR detectors have recently gained a lot of interest because of their potential to achieve extremely low noise equivalent temperature difference (NETD) while maintaining low cost to make them affordable to more applications. However, the curvature induced by residual strain mismatch within the microcantilever severely decreases the device performance. To meet performance and reliability requirement, it is important to fully understand the deformation of IR detectors. Therefore, the purpose of this study is threefold: (1) to develop an engineering approach to flatten IR detectors, (2) to model and predict the elastic deformation of IR detectors using finite element analysis (FEA), and (3) to study the inelastic deformation during isothermal holding.

INTRODUCTION

Bilayer microcantilever-based infrared radiation (IR) detectors have received extensive attention for wide use in military and civilian applications. These detectors can achieve a theoretical noise-equivalent temperature difference (NETD) of below 5 mK [1]. This type of IR detector is based on the bending of bilayer structures upon absorption of IR. The subsequent deformation can be readily determined by using piezoresistive, optical, or capacitive methods. However, the bilayer structures curve significantly after release from a sacrificial layer, largely due to the mismatch of residual stress/strain in the two materials [2]. Therefore, curvature modification is one of the important topics in the post-process assessment of IR detectors. It is also important to understand the deformation of IR detectors over a significant period of operation time, in order to meet performance and reliability requirements. The inelastic strain behavior (creep) in metal layers [3] results in inelastic deformation in IR detectors. Neglecting the inelastic deformation can lead to misinterpretations of the measurement data from IR detectors and can compromise performance. In previous study, the temperature and time - dependent deformations of SiN$_x$/Al bilayer microcantilever beams are characterized by using thermal cycling and isothermal holding testes [4]. In this study, we applied the same characterization methods on IR detectors to understand the thermomechanical behavior in device level. First, the thermal cycling technique was employed to flatten as-released IR detectors and characterize the linear thermoelastic behavior. Second, the characterized Power-law creep was used to develop a numerical model for predicting and simulating the inelastic behavior in long-term operation. The experimental methodologies and theoretical framework developed in this research can be readily applied to study the thermomechanical behavior of various bilayer microcantilever structures, and to improve the fundamental understanding for the design of microcantilever-based IR detectors.

EXPERIMENT

Specimen preparation

Figure 1 shows the fabrication process of IR detectors using a surface micromachining module with two sacrificial layers of polyimide. The use of spin-on polyimide allows not only an all-dry final release step overcoming stiction problems, but also complete compatibility with deposition and patterning of infrared structural layers, i.e., plasma-enhanced chemical vapor deposited (PECVD) silicon nitride (SiN_x) and electron beam (Ebeam) deposited aluminum (Al) in this work. The polyimide layer had a thickness of 2.5 μm and both the Al and SiN_x layers had a thickness of 200 nm. The anisotropic etching of polyimide was accomplished by high power reactive ion etching (RIE) using oxygen (O_2) at a rate of ~870 nm/min; the isotropic release etching was conducted by high density oxygen plasma asher at a rate of ~17.4 nm/min [5].

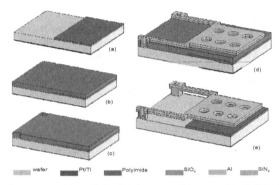

Figure 1: Fabrication process flow of the IR detector : (a) patterning Pt/Ti wires and pads by standard lift-off process, (b) coating and curing of 2.5μm thick polyimide, (c) deposition of SiO_x etching mask layer and etching of SiO_x and polyimide by using RIE with SF_6 and O_2 respectively, (d) deposition and patterning of the bimaterial cantilever of a 200 nm PECVD SiN_x layer and a 200 nm Ebeam Al layer, (e) release of the cantilever by using O_2 plasma asher.

The thermomechanical behaviors of the IR detectors were characterized by standard techniques, namely, thermal cycling and isothermal holding. Although these testing methods are less innovative, the standard procedures will generate more credible data, which are important for understanding the fundamental deformation mechanisms under different thermal loading histories.

Thermal cycling

In this study, the measurement procedure was designed to carefully explore the deformation of the MEMS-based IR detectors under uniform heating and cooling cycles. The heating and cooling cycles were performed over a range of temperatures, from room temperature to 265 °C - 310 °C. The MEMS-based IR detector had a heating rate of 200 °C/min. When the temperature approached the target temperature, the heating rate was decreased from 200 °C/min

112

to 10 °C/min, allowing the stage temperature to slowly approach the target temperature. The cooling rate was approximately the same as the heating rate for the MEMS-based IR detectors. Following the cooling period, the MEMS-based IR detectors were heated and subsequently cooled for four cycles, with the maximum temperature increase in each cycle set to 15 °C.

Isothermal holding

In order to study the effects of creep and stress relaxation at two different holding temperatures (100 °C and 125 °C), the deformation of the IR detectors was measured as a function of time. First, the MEMS-based IR detectors were thermal cycled three times between room temperature and 295 °C to partially stabilize the Al microstructure and diminish the initial curvature. After the initial pre-cycles, the MEMS-based IR detectors were held at the isothermal holding temperature (100 °C and 125 °C) for about 45 hours.

Measurement setup

A visual representation of the full-field out-of-plane deformation of microcantilevers with 100 nm resolution was achieved when a beam of white light passed through a 10× microscope objective to the surface of the microcantilever. Subsequently, an interferometer beam splitter reflected half of the incidence beam to the reference surface inside the microscope. These two beams were recombined and projected onto the camera to generate a signal that was proportional to the resultant beam intensity produced by the interference effect. These signals were then transferred into the spatial frequency domain, and the surface height for each point was obtained from the complex phase as a function of the frequency.

To induce the different temperature regions applied throughout the test, a thermal system, with a closed-loop temperature controller, a micro-heating/cooling stage and a cooling system (HCP302-STC200, INSTEC Inc.), was used. The resolution of the thermal system was approximately 1 °C.

RESULTS AND DISCUSSION

Planarity

The in situ thermal-mechanical response for a MEMS-based IR detector top bimaterial sensing plate is depicted in Figure 2. Upon return to room temperature, the curvatures after each thermal cycle are different. The difference is caused by the evolution of the residual stresses of the bilayer structures in the high temperature region (the nonlinear region is not plotted in Figure 2). For instance, the curvature at room temperature decreases from 7.3 mm^{-1} to 5.8 mm^{-1} after the first cycle, indicating a significant structure flattening. As shown in Figure 2, the curvature at room temperature continues to decrease when the MEMS-based IR detectors are subjected to a thermal cycling with a peak temperature higher than the previous peak temperature. In this study, the sensing plate initially bent downward with a curvature of 7.3 mm^{-1}. After the third thermal cycling with a peak temperature of 295 °C, upon return to room temperature, the curvature is found to be approximately to -0.1 mm^{-1}, a 97 % reduction from the initial value. Hence, the optimal annealing temperature should be between 280 °C and 295 °C, and the MEMS-based IR detector can be flattened by using thermal cycling with optimized peak temperatures. The result of this engineering approach is shown in Figure 3. Also, Figure 2 shows that the linear elastic response (*dk/dT*) of the IR detector before and after annealing at 295 °C are 0.053 mm⁻¹/°C and 0.055 mm⁻¹/°C, respectively. Therefore annealing doesn't affect the thermoelastic responses [5].

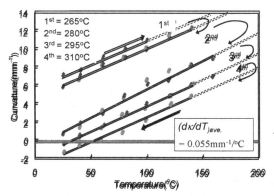

Figure 2. Curvature vs. temperature during the four thermal cycles with different peak temperatures. (Heating: blue symbols; cooling: pink symbols)

(a)   (b)

Figure 3. SEM image of IR detectors (a) before and (b) after thermal cycling with a peak temperature of 295 °C.

The as-fabricated MEMS-based IR detector bent up due to the residual strain mismatch from its two layers (Figure 4). Thermal cycling appears to be an effective method to modify the residual strain mismatch induced curvature. Figure 4 depicts the residual strain mismatch at room temperature after each thermal cycle from FEA. A uniform temperature loading was applied to each node in the finite element model. The magnitude of the temperature change was adjusted so that the simulated curvature of the microcantilever-based IR detector matched the measurement. The average strain mismatch along the central area of the sensing plate of the IR detector was then extracted from the FEA. The strain mismatch decreases with curvature reduction and the residual strain mismatch is zero when the microcantilever-based IR detector becomes flat.

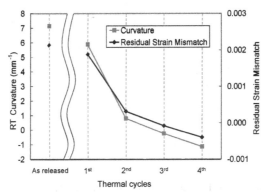

Figure 4. Measured room temperature (RT) curvature and calculated residual strain mismatch after each thermal cycle from Fig. 2.

Compared with the other known efficient flattening means for microcantilevers, such as ion machining [6] and rapid thermal annealing [2], thermal cycling is a nondestructive method and can also provide in-depth understandings on the thermal mechanical evolution of the device. When the top electrode has a particular curvature defined by the equation:

$$f(x) = \frac{1}{2}\kappa x^2 + bx + c$$

(1)

The capacitance (C_p) between two electrodes can be defined as

$$C_p = \varepsilon \int_0^\ell \frac{1}{d + f(x)} dx$$

(2)

where ℓ is the length of the top electrode, ε is the permittivity of the insulator and d is the gap between the two electrodes. The flattened MEMS-based IR detectors increase the capacitance signal readout.

Inelastic deformation

For the long-term isothermal holding, the SiN_x/Al microcantilevers were held at 100 °C and 125 °C for about 45 hours. The curvature of SiN_x/Al microcantilevers decreased by 1.6% and 15.1% of the initial curvature, respectively (Figure 5) due to the combined effect of creep and stress relaxation. FEA with Power-law creep ($\dot{\varepsilon} = A\sigma^n$) was used to describe the inelastic deformation behavior of the IR detectors during isothermal holding [5]. The microstructure evolutions due to isothermal holding in SiN_x/Al microcantilever beams were studied using an atomic force microscope in our previous study [5]. The microstructure evolutions associated with creep and stress relaxation can also be use to explain the deformation behavior of IR detectors in this study.

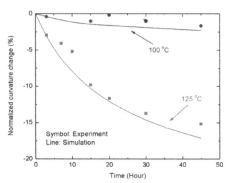

Figure 5. Experimental and simulation result of isothermal holding test on MEMS-based IR detectors.

CONCLUSIONS

In this work, thermal mechanisms were studied via thermal cycling to reduce the residual strain mismatch-induced curvature of the MEMS-based IR detectors. A thermal system and an interferometer were adopted to monitor the full-field curvature change in situ. Thermal cycling reduced the residual strain mismatch within the bilayer device and thus flattened the MEMS-based IR detectors. Although bilayer (SiN$_x$/Al) MEMS-based IR detectors were used in this design, this study is applicable to other MEMS-based IR detectors with different geometries and materials. Moreover, a FEA model with Power-law creep was developed for MEMS-based IR detectors, to predict the curvature evolution in short- and long-term operations.

ACKNOWLEDGMENTS

This project has been supported partially by the Young Faculty Award from DARPA/MTO to Dr. Y. Zhang (W911NF-07-1-0181), and the National Science Foundation through grant CMMI 0700688 and the Air Force Office of Scientific Research through grant FA 9550-06-1-0145 to Dr. X. Zhang. The authors would like to thank the Photonics Center at Boston University for all of the technical support throughout the course of this research.

REFERENCES

1. R. Amantea, L. A. Goodman, F. Pantuso, D. J. Sauer, M. Varghese, T. S. Villani and L.K. White, *Proceedings of SPIE*, **3436**, 647 (1998).
2. S. S. Huang, B. Li and X. Zhang, *Sens. Actuators A-Phys.* **130**, 331 (2006).
3. Y. Zhang, M. L. Dunn, K. Gall, J. W. Elam, S. M. George, *J. Appl. Phys.* **95**, 8216 (2004).
4. I.-K. Lin, X. Zhang, Y. Zhang, *J. Micromech. Microeng.* **19**, 085010 (2009).
5. I.-K. Lin, Y. Zhang, X. Zhang, *J. Micromech. Microeng.* **18**, 075012 (2008).
6. T. G. Bifano, H. T. Johnson, P. Bierden and R. K. Mali, *J. Micromech. Microeng.* **11**, 592 (2002).

Mater. Res. Soc. Symp. Proc. Vol. 1299 © 2011 Materials Research Society
DOI: 10.1557/opl.2011.59

Film Conductivity Controlled Variation of the Amplitude Distribution of High-temperature Resonators

Silja Schmidtchen, Denny Richter, Han Xia and Holger Fritze

Institute of Energy Research and Physical Technologies and Energy Research Center Niedersachsen, TU Clausthal, 38640 Goslar, Germany

ABSTRACT

High-temperature measurements of the spatial distribution of the displacement characteristics of a thickness shear mode langasite ($La_3Ga_5SiO_{14}$) resonator are obtained using a laser Doppler interferometer. Thereby, the resonator is excited in the fundamental mode and the third overtone. Further, the resonator is coated with a gas sensitive CeO_{2-x} film which exceeds the metal electrode. In reducing atmospheres the conductivity of the film increases and induces an increase of the effective electrode area. This effect leads to a broadening of the mechanical displacement distribution. The latter depends strongly on the size of the excited part of the resonator which is determined by the effective size of the electrodes. The direct determination of the mechanical displacement at different oxygen partial pressures confirms a model as derived from the electrical impedance of resonator devices [1]. Further, information about the mass sensitivity distribution of resonators is obtained since the property is directly proportional to the amplitude.

INTRODUCTION

High-temperature stable langasite resonators can be used to determine atmosphere-induced changes in the mechanical and electrical properties of metal oxide films at elevated temperatures [1]. Potential devices include gas sensors for high-temperature applications showing an increased selectivity to different reducing gases like H_2 and CO in comparison to conventional conductivity based metal oxide sensors [2]. Alternatively, piezoelectric transducers can be used to study the electrical and visco-elastic properties of metal oxides under different conditions [3]. In the presented work, the principles of the conductivity-induced variation of the effective electrode diameter and its influence on the distribution of the mechanical displacement of piezoelectric resonators are determined and discussed.

For this purpose a CeO_{2-x} film is applied on top of the metal electrode of the resonator. Since the area of the CeO_{2-x} film is larger than the metal area, a change in the conductivity of the film causes a change of the effective electrode area and therefore of the excited area of the resonator. The spatial distribution of the displacement amplitude and, thereby, of the mass sensitivity depends on the size of the excited part of the resonator as shown schematically in FIGURE 1.

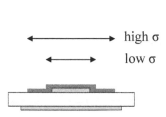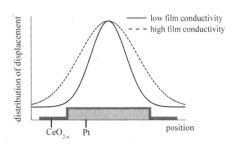

FIGURE 1: Schematic drawing of a resonator with platinum electrodes and an additional metal oxide film. With increasing conductivity, the effective electrode area increases. The arrows indicate the cases of insulating CeO_{2-x} film (low σ) and high conductivity (high σ) (left). Distribution of displacement depending on the film conductivity (right).

In this configuration, the frequency shift is not primarily caused by changes in mass load, but rather by changes in the mass sensitivity distribution of the resonator. The distribution of the sensitivity of a thickness shear mode resonator can be approximated by a Gaussian function having the maximum at the center of the electrode [4]. The width of the Gaussian profile depends predominantly on the diameter of the conductive area. Increasing the conductivity of the metal oxide film increases the effective electrode diameter for the configuration shown in FIGURE 1 and broadens the sensitivity distribution. This leads to an increase in sensitivity in the area of the relatively heavy platinum electrode. Measurements using a laser Doppler vibrometer are performed at high temperatures to determine the shear mode amplitude in order to verify this effect.

EXPERIMENTAL

Langasite is a single crystalline piezoelectric material that exhibits a very good thermal and chemical stability. It can be operated piezoelectrically up to temperatures of about 1400 °C [5]. The bulk acoustic wave resonator used in this work is a y-cut langasite disc with 19 mm in diameter and a thickness of 600 μm. The resonance frequency of the fundamental thickness shear mode is 2.2 MHz. The electrodes are keyhole shaped platinum films of different diameter at front and rear side. The smaller electrode, 5.8 mm in diameter, is deposited by pulsed laser deposition and shows a thickness of 300 nm. The larger electrode, 10.3 mm in diameter, is screen printed to ensure a sufficiently rough surface since a diffuse scattering is required for the vibrometer measurements. The thickness of this electrode is about 3 μm. The metal oxide film is a pulsed laser deposited CeO_{2-x} film with a thickness of 200 nm. It is applied on the smaller electrode and shows a diameter of 13.2 mm thereby overlapping the platinum electrode (see FIGURE 1).

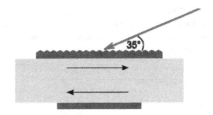

FIGURE 2: Schematic sketch of the laser spots for linear scanning. Dotted line symbolizes the platinum electrode at the rear side of the resonator (left). The arrows shown inside the resonator indicate the direction of the shear vibration (right).

Measurements are performed at room temperature or in a gas tight tube furnace operated at 580 °C. The different oxygen partial pressures are adjusted using a gas mixing system and measured with a zirconium oxide based oxygen sensor. A laser Doppler vibrometer (Polytec OFV 505) is placed outside the furnace at a distance to the sample surface of about 0.5 m. The minimum achieved resolution of the displacement is at least 100 pm. The sample surface is scanned linearly with a step size of 250 μm in the direction of the vibration (FIGURE 2). The angle of incidence is 35° which requires a correction of the measured displacement in order to get the displacement of the shear movement. For three different oxygen partial pressures, i. e. 10^{-4}, 10^{-16} and 10^{-20} bar, the resonator surface is scanned at 50 different positions.

The oxygen partial pressure dependent conductivity of the CeO_{2-x} film is measured independently using an impedance analyzer. The bulk properties of the resonators such as the bulk resistance R_b are investigated by impedance spectroscopy at frequencies much below the resonance frequency with an impedance analyzer Solartron 1260. The resonance properties are determined in the vicinity of the resonance frequency with a network analyzer HPE5100A [1].

DISCUSSION

The distribution of the displacement depends on the position on the resonator as shown in FIGURE 3 for room temperature. Again, the sample is excited with the resonance frequencies of the fundamental mode and the third overtone.

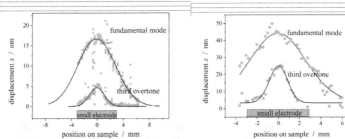

FIGURE 3: Spatial distribution of displacement (zero-to-peak) of a 2 MHz resonator, excited at its resonance frequencies of the fundamental mode and third overtone. The excitation voltages are $U_A = 2$ V for the fundamental mode and 5 V for the overtone. The measurements are performed in air at room temperature (left). Spatial distribution of the displacement measured at 10^{-16} bar at 580 °C. The excitation voltage is $U_A = 9$ V for both (right).

The results show a significantly wider profile for the fundamental mode than for the overtone. Further, the distribution at room temperature shows about the same width than that at 580 °C. The fact becomes obvious by comparing data normalized with respect to the maximum amplitude (not shown).

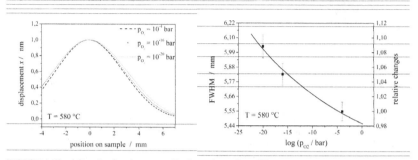

FIGURE 4: Fitted Gaussian functions, normalized to unity, at three different oxygen partial pressures (left). Full width at half maximum (FWHM) of fitted Gaussian functions at different oxygen partial pressures (right).

A similar behavior is observed for measurements in reducing atmospheres at a temperature of 580 °C. Again, the results are fitted with Gaussian functions. The fits are displayed in FIGURE 4 (left). The data show a broadening of the vibration profile with decreasing partial pressure, which corresponds to an enlargement of the effective electrode area. Calculating the full width of half maximum it becomes obvious that the reduced oxygen partial pressure leads to an increase in the effective electrode area by 18 % which corresponds to an increase in diameter by a factor of 1.09.

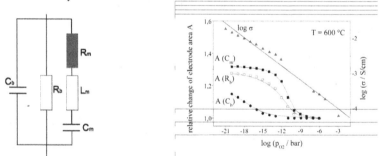

FIGURE 5: Extended Butterworth-van-Dyke equivalent circuit (left). Relative changes of the parameters of the electrical equivalent circuit which are related to the oxygen partial pressure dependent effective electrode area. Oxygen partial pressure depending conductivity of the CeO_{2-x} film (right).

The broadening of the vibration distribution follows indirectly from electrical measurements as described earlier in [1]. The resonator material langasite does not show changes in conductivity at 580 °C if the oxygen partial pressure is lowered down to 10^{-21} bar. Therefore, changes in electrical properties of a resonator configuration as shown in FIGURE 1 must be related to the sensor film, in particular, to the extent of the electrode. In order to quantify the latter impact, the electrical behavior of resonator devices is approximated by the extended Butterworth-van-Dyke equivalent circuit (BvD-EC, FIGURE 5, left) [6]. It includes bulk and motional properties denoted by the index b and m, respectively. They show a distinct dependence on the electrode area, i. e. capacitances C are proportional to the electrode area A while resistances R are inversely proportional. FIGURE 5 (right, left axis) depicts the oxygen partial pressure dependent increase of the area related to the above mentioned BvD-EC parameters. They show clearly the same trend while the behavior of bulk and motional differs slightly. Detailed modeling describes the behavior and difference properly. Having in mind that the conductivity of CeO_{2-x} increases with decreasing oxygen partial pressure as also seen in FIGURE 5 (right, right axis) the changes of the BvD-EC parameters can be clearly attributed to the oxygen partial pressure dependent extent of the electrode conductivity. For example, the increase of the motional capacity C_m corresponds to an increase of the effective electrode area by about 15 %. This corresponds to an increase of the effective electrode diameter by a factor 1.07. This value is close to that obtained by laser Doppler vibrometry as described above.

CONCLUSIONS

Measurements of the vibration profiles of a langasite resonator are performed successfully at room temperature and 580 °C using a laser Doppler interferometer. The comparisons of the fundamental mode and the third overtone show a significant smaller displacement profile of the third overtone compared to the fundamental mode. The fact must be regarded while calculating the mass sensitivity of such resonators.

Further, the impact of the electrode conductivity on the spatial distribution of the displacement is demonstrated clearly. The results are in accordance with electrical measurements which indicate an impact of the electrode conductivity on the resonating area. The knowledge of the oxygen partial pressure dependent variation of the displacement profile of a metal oxide film coated langasite resonator enables to design improved high-temperature gas sensors.

ACKNOWLEDGMENTS

Financial support from the German Research Foundation made this work possible.

REFERENCES

1. H. Fritze, D. Richter, and H. Tuller. Simultaneous detection of atmosphere induced mass and conductivity variations using high temperature resonant sensors. *Sensors & Actuators: B. Chemical*, **111**:112, (2005).
2. D. Richter, T. Schneider, S. Doerner, H. Fritze, P. Hauptmann. Selective Gas Sensor System for CO and H_2 Distinction at High Temperatures Based on a Langasite Resonator Array. *Solid-State Sensors, Actuators and Microsystems Conference, TRANSDUCERS 2007, International* (2007) 991-994
3. R. Lucklum and P. Hauptmann. The quartz crystal microbalance: mass sensitivity, viscoelasticity and acoustic amplification. *Sensors & Actuators: B. Chemical*, **70**(1):30–36, (2000).
4. B. Martin and H. Hager, Velocity profile on quartz crystals oscillating in liquids. *J. Appl. Phys.* **65**, 2630 (1989)

5. J. Sauerwald, H. Fritze, E. Ansorge, S. Schimpf, S. Hirsch, B. Schmidt, Electromechanical properties of langasite structures at high temperatures, *International Workshop on Integrated Electroceramic Functional Structures, Berchtesgaden, Germany, 6.-8.6. (2005).*
6. H. Fritze. High-temperature bulk acoustic wave sensors. *Measurement Science and Technology, Meas. Sci. Technol.* **22** () 012002 (28pp), (2011).

Mater. Res. Soc. Symp. Proc. Vol. 1299 © 2011 Materials Research Society
DOI: 10.1557/opl.2011.60

Ultrafine Silicon Nano-wall Hollow Needles and Applications in Inclination Sensor and Gas Transport

Z. Sanaee, S. Mohajerzadeh, M. Mehran and M. Araghchini

Nanoelectronic Center of Excellence, Thin Film and Nano-Electronic Laboratory, School of

Electrical & Computer Engineering, University of Tehran, Tehran, Iran

ABSTRACT

We report on the realization of high precision hollow structures directly on silicon suitable for liquid and gas/vapor transport. The formation of hollow structures requires high aspect ratio etching combined with bulk back-side micro-machining to realize silicon-based membranes. The use of a slant angle deposition method has been used as an alternative method for three-dimensional lithography. The transfer of acetone vapor through such tiny holes shows an anomalous behavior where a sharp rise is observed followed by an exponential and gradual decay. These structures can be eventually used as mass ion separation devices.

INTRODUCTION

Smart drug delivery depends on the evolution of hollow micro-needles [1]. The transfer of drug through little holes inside such structures allows a replacement for regular injection needles [2, 3]. In addition, the contact between the electrodes and the patient's body in electrocardiograph units is usually made through a gel surface. Micro-needles are considered as possible alternatives to replace the electrolytic gel, where they can penetrate through the outer layer of skin and improve the electrical contacts with no need to an externally applied gel [4]. Apart from biological applications, the cup-like structures can be used as gas and vapor transfer media for mass spectroscopy and ionization sources [5]. We report a method to realize ultrafine hollow needle structures on silicon-based membranes. Such structures were used for liquid and gas transport showing anomalous behavior.

EXPERIMENT

High aspect ratio deep etching of silicon is used to fabricate hollow micro-needle. Deep etching of Si has been practiced using a modified reactive ion etching method where a combination of $H_2/O_2/SF_6$ gases is used to perform passivation and SF_6 gas for etching step in fully programmable sub-cycles and in a sequential manner. These passivation/ etching sub-cycles should be repeated as many times to achieve desired depth. Details about this technique can be found elsewhere [6].

The fabrication starts by cleaning (100)-Si wafers followed by E-beam deposition of Cr as the mask for future processing. Circular hollow rings with outer diameter of 3-40 μm and rim widths less than 200 nm are created using projection lithography. Once vertical pipe-like structures are created, a slant-angle rotation and deposition method is used to cover the surface of the samples with Cr except for the inner surfaces. As a result only the outer sides of the recessed craters are coated while inner parts and especially the inner bottom remains uncoated. Since the bottom of each micro cylinder is not coated with Cr, the etching can be continued on these surfaces without affecting the other parts of the silicon surface. To obtain through-holes, the Si surface needs to be thinned prior to etching, accomplished using backside micromachining in KOH solution. For this purpose standard KOH solution with 30% molar concentration is used at a typical temperature of 60°C for 25 hours. Typical thicknesses of the membranes of the order of 30-40 μm are found to be suitable for this process. Using this method, Si micro-needles with outer diameter size from 3-40 μm are fabricated. High-resolution vertical etching gives the opportunity to have needle wall thicknesses about 100 nm where the height of each feature can be as high as 20 μm. Aspect ratios in excess of 60 are easily obtained, although higher values are also achievable. Figure 1 shows the fabrication process.

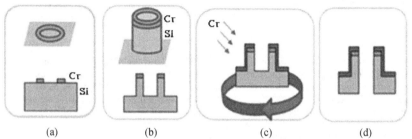

(a)　　　　　　　(b)　　　　　　　(c)　　　　　　　(d)

Figure 1: The formation of hollow structures using (a) ring patterning, and (b) deep reactive ion etching, followed by, (c) small angle deposition and (d) second deep reactive ion etching.

Figure 2 collects some of the SEM images of the pipe-like structures with high precision features.

(a)　　　　　　　　　　(b)　　　　　　　　　　(c)

Figure 2: A collection of several deeply etched hollow micro-needles, where aspect ratios of more than 60 has been achieved. (a) and (b) array of 8um micro-needles (c) array of 18um micro-needles.

The creation of through-holes in these structures can be verified using optical microscopy, where some of the images are presented in Figure 3. This experiment has been conducted with a back-light illumination to observe the evolution of holes in the silicon membranes. Part (c) in this figure shows the SEM image of backside of a sample where through-holes can be distinguished.

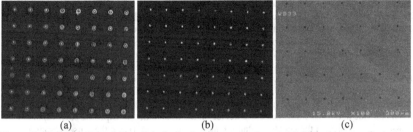

(a)   (b)   (c)

Figure 3: The evolution of through-holes in the silicon membrane. (a) Optical image of micro-needle sample while both front and back light is on. (b) By turning off the front light, the back-light passing through the holes shows the evolution of holes through the silicon membrane (c) SEM image of back side of sample further shows hollow micro-needles are successfully fabricated.

DISCUSSION

These structures have applications as inclination sensors based on a liquid (water) penetration through their holes. The presence of a cavity underneath the silicon membrane (realized at the back-side etching step) was found to be suitable to encapsulate a droplet of water. The structure of the capacitor is shown schematically in Figure 4 where the backside of silicon micro-needle sample acts as one electrode of the capacitance whereas a chromium-plated glass on top acts as the second electrode. By penetration of water into the tiny holes of micro-needle, the value of the capacitance can be significantly altered, thanks to the high value of permittivity of water as opposed to that of silicon and air. Part (b) in Figure 4 shows the measured value of the capacitance as a result of water inclusion, after water inclusion and after tilting the sample with a small angle of 5°C.

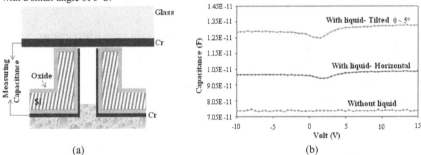

(a)   (b)

Figure 4: (a) The schematic representation of the liquid-included capacitance and (b) the measurement results of dry and liquid-included sample before and after tilting.

The value of the capacitance reads 8 pF for the bare capacitor and rises to 10 pF after water inclusion. A second increase to 13 pF is observed after the whole structure is tilted where the level of water penetration is increased. The measurement of the capacitance-angle has been repeated to observe the reversibility of the data where similar values are recorded for inclination angles below 10 degrees. A theoretical modeling for this behavior is being developed.

Such structures can also be used as media to transfer gas or vapor through their tiny holes for which we have used a setup as shown in Figure 5. A large reservoir is used to hold the vapor inside with a uniform distribution. We have used tin-oxide semiconducting oxide gas sensors (SnO_2) fabricated in-house to detect the trace of gases and vapors used for this investigation. SnO_2 sensors are sensitive devices, which detect various hydrogen-based and hydrocarbon gases and vapors, as acetone and alcohol vapors. To observe the transient behavior of the hollow structures one can place the needle-holding silicon membrane between the gas/vapor containing bottle on one side and a gas/vapor-detecting device on the other side. By exposing such vapors to the sensing device through the needle-containing membrane, a sharp rise in the sensor response is observed, which could be due to an enhanced transport due to surface effects. The pressure in the reservoir and the sample side remain at ambient pressure.

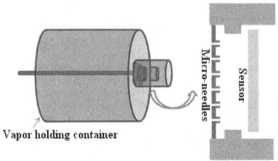

Figure 5: The schematic setup used for detection of vapor transport through the little holes of the micro-needles. The sensing device in this experimental setup is a tin-oxide semiconducting oxide detector.

Figure 6 collects the results of this investigation on acetone at different vapor concentrations with and without the presence of a silicon membrane. As observed from these figures by using a micro-needle structure, a sharp rise is observed in the sensor signal, followed by a gradual decay. This behavior could be due to the surface enhanced transient of vapor molecules through the tiny holes of silicon-based needles. Further investigation on the understanding and modeling of this anomalous behavior is underway.

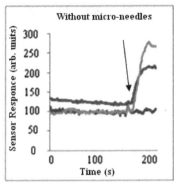 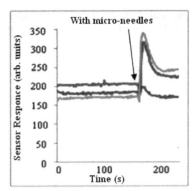

Figure 6: Sensor response to acetone vapor. Different colors correspond to various concentrations. As seen in this figure, for the case where micro-needles are placed in the path of vapor to the detector, a sharp rise is observed in the sensor signal, followed by a gradual decay. Arrow in each figure shows the gas-in moment for each experiment. In both figures, the green, red and blue curves refer to approximate concentrations of 1000, 500 and 100 ppm of acetone, respectively.

CONCLUSIONS

We have successfully fabricated and tested silicon-based micro-needles with high aspect ratios and ultra fine features. The formation of such structures has been possible using a deep reactive ion etching combined with small angle deposition method to allow the removal of silicon from desired places inside the cylindrical structures. The tiny hallow needles have been exploited for the formation of liquid-included capacitors as well as vapor/gas transport. The presence of such sieve-like membranes can be used for ion mass separation and mass spectroscopy. Further investigation on the etching process as well as gas/vapor transport properties of these structures is being pursued.

REFERENCES

1. M. R. Prausnitz, Advanced Drug Delivery Reviews 56, 581 (2004).
2. P. Khanna, J. A. Strom, J. I. Malone and S. Bhansali, Journal of Diabetes Science and Technology2, 1122 (2008).
3. G. Holman, Y. Hanein, R. C. Wyeth, A. O. D. Willows, D. D. Denton and K. F. Böhringer, 2nd Annual International IEEE-EMBS special Topic Conference on Microtechnologies in Medicine and Biology, 225 (2002).
4. L.M. Yu, F.E.H. Tay, D.G. Guo, L. Xu and K.L. Yap, Sensors and Actuators A 151, 17 (2009).
5. L. Y. Chen, L. F. Velásquez-García, X. Wang, K. Teo and A. I. Akinwande, Technical Digest - International Electron Devices Meeting, IEDM, 843 (2007).

6. A. Sammak, S. Azimi, N. Izadi, B. Khadem Hosseinieh, and S. Mohajerzadeh, IEEE Journal of Microelectromechanical systems 16, pp. 912 (2007).

Mater. Res. Soc. Symp. Proc. Vol. 1299 © 2011 Materials Research Society
DOI: 10.1557/opl.2011.61

Development of a Robust Design of a Wet Etched Co-integrated Pressure Sensor

Wolfgang Schreiber-Prillwitz[1, 2], Mikko Saukoski[2], Gerhard Chmiel[2], and Reinhart Job[1]
[1]University of Hagen, Dept. of Mathematics and Computer Science, 58084 Hagen, Germany
[2] ELMOS Semiconductor AG, 44227 Dortmund, Germany

ABSTRACT

The performance of a co-integrated silicon pressure sensor for the 1-bar full scale range was optimized. A gain in signal of ca. 5% was calculated and verified by optimizing the piezoresistors position on the membrane. The influence of alignment errors between the backside cavity mask and the positions of the piezoresistors on the membrane's front side were calculated. Depending on the asymmetry, a maximal electrical signal deviation of 1% was found. The impact of underetching effects (KOH) at the backside mask on electrical signals was also analyzed. Underetching has a certain range, alters the membrane size, and has a strong impact on sensor performances. In a worst case scenario signal variations caused by underetching could be finally reduced from 15% to 4%.

INTRODUCTION

The requirements of the performance of pressure sensors are increasing by the rapidly growing applications within medical, industrial, and mainly automotive areas. Certain constraints in the latter field are strictly defined procedures, which confine the options of possible solutions. Beside long term stability and overall accuracy, the uniformity of certain key parameters, e.g. sensitivity of pressure sensors, is demanded. In general, a small variation of any parameter, caused by inevitable process deviations in mass production, guarantees high yield, ability to deliver, and cost effective production. A common demand for acceptable parameter distribution is the 6 σ level, which defines the tolerance limit for a parameter as six times the standard deviation from the average value of its distribution. This results in a failure rate of only 3.4 per million.

For quality and space saving reasons, the read out circuitry and the pressure sensitive structure are often combined as a monolithically integrated sensor. The benefit concerning quality is that no wire bonding between both system components is required, and therefore, no failure due to breaking bond wires can occur. Space wise, the need of bond pads for the sensor cell and a certain frame around the sensing membrane only for assembly is omitted: the area around the membrane can be used for circuitry. This monolithic approach is a system consisting of two components. The circuitry typically is realized by a CMOS process. CMOS technology follows very strict rules and process step sequences, which need to be qualified and released for series production in automotive applications, following exactly defined process parameters and order. Hence, also the sensing element as the second component has to be designed within this framework of primitive devices (resistors, capacitors, transistors, etc.) and design rules of the CMOS process. The challenge in case of a monolithic integrated pressure sensor system is to enable an optimized sensor design within the released library of primitive devices of the CMOS process.

In [1], we have shown and verified a design flow that leads from structural finite element analysis (FEA) of a silicon membrane to electrical signals of a Wheatstone Bridge. For an existing co-integrated pressure sensor system for a full scale pressure range of 1 bar, in this paper we applied this method to investigate the optimization potential of the design to get a maximal FS signal, with minimal sensitivity deviation due to process related effects like underetching and front-to-back misalignment, and focused on process uncertainties of the membrane. By etching the silicon from the backside of a wafer to form a membrane, the membrane defining mask will undergo a certain amount of underetching. Underetching describes, how far KOH will go laterally underneath the passivation of the backside membrane mask, and so increase the membrane opening compared to the mask opening. Process wise, this effect mainly depends on the stability and the quality of the back side preparation of the wafer surface (polishing), and the quality of the mask layer deposition (e.g. deposition of the oxide/nitride layer stack). A certain underetching is normal; typical underetching rates are in the range of several 10 μm, and are considered in the design of the backside mask. The consequence of a change of the underetching rate would be an altering membrane size and thus an altering distance between the membrane rim and the piezoresistive bridge resistors. As the mechanical stress levels on membranes have a steep slope near the rim, relatively small differences of the distances of the piezoresistors to the rim would result in significant different signals or sensitivity of the sensor. The misalignment describes the front-to-back-positioning failure of the backside mask for the membrane opening towards the front side resistor bridge configuration. This results in an asymmetrically shift of the resistor bridge configuration to the membrane rim, and is also influencing the output signal of the sensor.

METHOD

Starting point is a 3-D geometric model of the sensor structure. In contrary to the often used approach of only modeling one half or even one quarter of the structure, utilizing symmetries of the device, a full model was simulated. This enables us to investigate the asymmetric configuration of the resistors within the membrane as it occurs in the case of misalignment. On the other side, a simplification was made by neglecting the passivation layer compound. For this optimization run, the oxide/nitride layer stack of the passivation of the membrane was not considered, as only the relative change of the signal was of interest, and not the absolute value. The thin layers of the passivation (below 1 μm thickness), together with wide lateral dimensions in the range of 1000 μm, would need a high number of elements within the FEA calculation. The effect of neglecting the oxide/nitride layer stack for signal evaluation is an increase of the signal of ca. 20% compared to the real devices with 13.2 mV Fs signal. Signal evaluations starts with structural simulations of the deflection of the silicon membrane under pressure load. For this purpose the FEA tool ANSYS® is used, which delivers membrane deflection, normal stress values σ_{xx}, σ_{yy}, and the shear stress σ_{xy}. According to Tufte [2], the relation between resistivity change of implanted resistors in silicon and applied physical stress can be expressed by

$$\frac{R}{R_0} = \pi_l \cdot \sigma_l + \pi_t \cdot \sigma_t \qquad (1)$$

where σ_l and σ_t are the longitudinal (current and stress are parallel) and transverse (current and stress are perpendicular) stress values along the according resistors, and π_l and π_t are the piezoresistive coefficients, respectively. In the used coordinate system, the resistors are oriented with their length parallel to the x-axis, and their width along the y-axis, respectively. Hence, the normal stress σ_{xx} equals σ_l, and the normal stress σ_{yy} equals σ_t. A comprehensive derivation of the piezoresistive effect is given in [3]. The conversion of stress distributions to electrical signals was done by a MATLAB® program, which interpolates the nodal stress values from FEM simulation and matches them to the resistor geometries and locations. The routines take into account process related parameters like p-n junction depth, or doping profiles. Resistivity changes for each resistor are calculated from the difference of the stress values for the unloaded membrane (zero pressure) and the full scale pressure (1 bar). The experimental verification of this method is described in [1].

Applying this signal evaluation method towards the existing design of an 1-bar (14.5 psi) pressure sensor without passivation layer stack, a FS-signal of 16.8 mV/V load was calculated. To investigate the characteristics of the FS signal around the nominal position of the resistors, their coordinates within the evaluation routines are modified, and the bridge signal for different coordinates on the membrane are determined. The goal is to check if the actual position of the resistors (chosen by best practice) is really delivering the highest signal. Coming from the target positions, the resistor coordinates for all four resistors are simultaneously shifted in direction of the membrane center, then in the opposite direction towards the rim. In this paper, this is denoted as 'radial displacement'. The calculated bridge signal for a certain pressure load will be highest at the optimal position of each resistor on the membrane.

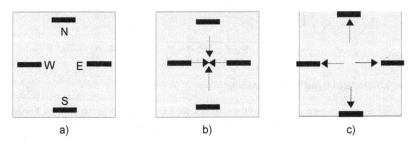

Figure 1. Underetch: a) nominal, b) stronger (radial inwards), and c) less (radial outwards).

To investigate the influence of underetching, the same method as for the position optimization is used. The start position of the resistors is symmetrically with respect to the membrane edges (Fig. 1a). The coordinates of the resistors are then changed symmetrically with regard to the midpoint of the membrane (radial shift). As mentioned before, there is always certain underetch. In case of an increasing underetch rate, the membrane would become wider than targeted, which leads to a resistor position relatively shifted inwards to the membrane center compared with the target position. Within the signal evaluation script, this effect is realized by changing the resistor coordinates towards the midpoint of the membrane (Fig. 1b), and vice versa (Fig. 1c).

RESULTS AND DISCUSSIONS

Fig. 2 shows the calculated FS bridge signal vs. the radial displacement for the existing design. '0' on the x-axis represents the resistor locations on the membrane as they are on the original design. Negative x-values shift all resistors inwards towards the membrane center, positive outwards towards the rim. It can be seen, that higher signals can be obtained by placing the resistors more towards the membrane rim, and that the optimal position was not realized yet. Tab. 1 shows the FS signal change in percent relative to the actual position. A radial shift between +10 μm to +15 μm (outwards) compared to the actual position will deliver a gain in sensitivity of ca. 5%. As explained above, a deviation of the underetching rate can be interpreted as a simultaneous symmetrical radial movement of the resistors inwards (higher rate) or outwards (lower rate) on the membrane, compared to the zero position. Tab. 1 also shows that for the actual design a change of the underetching rate of ±10 μm (represented by 0 ± 10 μm) would lead to a sensitivity change of the device of ca. 15% (−10% to +5.4%). Hence, a new resistor target position will be investigated at 13 μm outwards compared to the original position, between 10 μm and 15 μm.

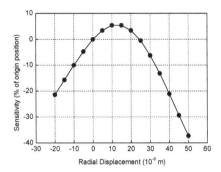

Figure 2. Calculated signal vs. resistor shift.

Based on the new target position of the resistors, the same routine was run through. For investigation of the effect of underetching, the coordinates were swept around the target point. The misalignment is mainly due to the front-to-back-positioning failure of the backside mask for membrane opening (Figs. 3a – c). The result would be an unsymmetrical position of the resistors towards the x-axis, y-axis, or a diagonal shift. A maximal misalignment of ±5 μm is reasonable from the process. The calculation was done from −15 μm to +15 μm to look at the effect beyond the limits (Tab. 2). The worst case would be the diagonal shift in Fig. 3c), as the symmetry of the bridge towards the membrane is disturbed most. Therefore, only this case is looked upon.

Table 1. Sensitivity change relative to the actual resistor position (0).

Radial Move-ment x and y [μm]	Delta Signal [%]
-20	-21,4
-15	-15,7
-10	-10,0
-5	-4,7
0	0,0
5	3,4
10	5,4
15	5,4
20	3,4

Signal gain ~5%

Underetching Δ ~15%

Fig. 4 shows the signal change in % FS vs. radial displacement, compared to the new target position. Now, any deviation from the target value results in a decrease in sensitivity. An underetching rate of ±10 μm for the new configuration results in a FS signal deviation of ca. −4%

(Tab. 2). From Tab. 3 it can be seen, that the signal loss due to misalignment is reduced for the not optimized resistor position compared to the optimized one. The reason is that for the old design any shift of the resistors does not move each resistor from its optimal position, but for the diagonal shift always two resistors are moved more outwards, and so will deliver higher signals. On the contrary, at the sensitivity optimized locations, all resistors always will lose signal. In the range of about 5 μm in x-direction or y-direction, the difference is quite small and negligible compared to the signal gain and less signal deviation due to potential underetch effects.

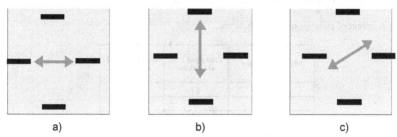

a)   b)   c)

Figure 3. Effect of misalignment.

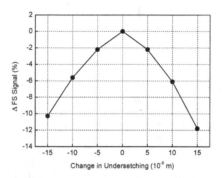

Figure 4. Signal change in % FS versus radial displacement (new target position).

Table 2. Sensitivity change relative to the optimized resistor position.

Radial Movement x and y [μm]	Delta Signal [%]
-15	-7,8
-10	-3,6
-5	-0,9
0	0,0
5	-0,9
10	-3,6
15	-7,8

Underetching Δ ~4%

There are two corners to look at, i.e. the +7 μm (over the limits) misalignment in x- and y-direction (worst case), and the +10 μm underetch (membrane is 20 μm wider than target). This will lead for one half of the bridge (N-E) to lower stress values compared to the target position, as the movement of the membrane edge (10 μm outwards) is more than the +7 μm of the misalignment. Even more, this is valid for the other half (S-W) of the bridge, as the resistors move relative towards the mid of the membrane, and so away from the stress maximum. For the worst case scenario with a +10 μm underetch and +7 μm diagonal shift – corresponding to a misalignment in x- and y-direction of +5 μm – the delta FS signal is –6%.

Table 3. Effect of misalignment on FS signal.

Diagonal shift [μm]	delta signal [%] old position	delta signal [%] new position
-15	-6,1	-9,0
-10	-2,7	-4,2
-5	-0,7	-1,1
0	0,0	0,0
5	-0,8	-1,1
10	-2,8	-4,2
15	-6,2	-9,0

CONCLUSIONS

Based on a proven methodology of evaluating signals from piezoresistors in a Wheatstone Bridge configuration within a stress distribution, a pressure sensor device optimization concerning process deviations was carried out. It was found, that a new position 13 μm closer to the membrane rim compared to an existing position will be best to improve sensitivity as well as sensitivity distribution caused by typical process deviations (underetching and front-to-back misalignment) in volume production. Beside a possible gain in full scale signal of 5% compared to the old design, the worst case deviation range in sensitivity due to process deviations can be decreased from 16% to 6%. In the context of parameter distribution this is important, as the upper and lower tolerance limits for the 6 σ level can drastically be narrowed.

REFERENCES

1. W. Schreiber-Prillwitz, M. Saukoski, G. Chmiel, R. Job, ECS Transactions, **33(8)**, 327 (2010)
2. O. N. Tufte, P. W. Chapman, D. Long, J. Appl. Phys., **33**, 3324 (1962)
3. E. V. Thomsen, J. Richter, Piezo Resistive MEMS Devices: Theory and Applications, http://www2.mic.dtu.dk/research/prototyping/piezo/piezonote2005_ver7.pdf

Mater. Res. Soc. Symp. Proc. Vol. 1299 © 2011 Materials Research Society
DOI: 10.1557/opl.2011.57

SU8 / modified MWNT composite for piezoresistive sensor application

Prasenjit Ray[1], V. Seena[1], Rupesh A. Khare[2], Arup R. Bhattacharyya[2], Prakash R. Apte[1], Ramgopal Rao[1]
1. Centre of Excellence in Nanoelectronics, Department of Electrical Engineering, Indian Institute of Technology Bombay, Mumbai, Maharashtra, India.
2. Department of Metallurgical Engineering & Materials Science, Indian Institute of Technology Bombay, Mumbai, Maharashtra, India.

ABSTRACT

SU-8 is being increasingly used as a compliant structural material for MEMS applications due to its interesting properties such as lower Young's modulus and higher mechanical and thermal stability. One of the popular classes of MEMS devices is a piezoresitive microcantilever. Ultra-sensitive polymer composite cantilevers made up of SU-8 as a structural layer and 10% carbon Black in SU8 as a piezoresistive layer with lower Young's modulus and higher gauge factor have been reported recently by our group. Higher electrical conductivity at lower concentration of conductive filler is of increased interest. Here we report a novel composite with purified multiwall carbon nanotubes (MWNT) in SU8 as a piezoresistor. MWNT were modified with octadecyl triphenyl phosphonium bromide (OTPB) in order to achieve debundled MWNT. A microcantilever device with integrated MWNT/SU-8 composite has been fabricated and characterized.

INTRODUCTION

Microcantilevers are normally used in Atomic Force Microscopy (AFM) imaging, for bio-molecular sensing and for applications in chemical sensing. For these applications optical detection (using laser) is used for sensing the deflection of the cantilever beam. However, optical detection scheme is rather complex and laser alignment makes this sensor inconvenient to use in the field. Due to this drawback, the integrated piezoresistor scheme looks very attractive. Several integrated piezoresistive cantilevers have been developed for AFM imaging [1-3], environmental sensors [4], bio-sensors [5-6] and mass-sensors [7]. Though most of the microcantilevers have been silicon based, only recently polymer based sensors based on materials such as SU8, because of their lower Young's modulus and ease of fabricating high aspect ratio structures, have become more attractive alternatives. Earlier, SU8 based polymer cantilevers with gold (Au) and polysilicon have been tried as piezoresistive layers [8-10]. Due to the lower gauge factor of gold, of the order of 2, the gold strain gauge based cantilevers are not suitable for practical applications such as bio-sensors. Carbon black/SU8 composite is reported recently as another option for piezoresistive polymer cantilever with a very high gauge factor of 15-20 with 10 wt% CB in SU8 [11]. These cantilevers have been used successfully for bio applications and explosive detection application [12, 13]. Instead of carbon black, the use of multiwall carbon nano tubes (MWNT) is expected to show improvement in dispersion and hence resulting in higher gauge factor for MWNT/SU8 composite compared to CB/SU8 composite. In this paper we report fabrication and characterization of piezoresistive microcantilevers using SU8/MWNT composite, with two different concentration of MWNT (0.1 wt% and 0.2 wt %).

EXPERIMENTAL DETAILS

Flip-chip technique [11]-[12] is used to fabricate piezoresistive cantilever with SU8/MWNT. It consists of five levels of lithography steps. The process starts with an oxidized wafer with thickness of 1000 nm. This oxide will act as a sacrificial layer at the last stage of the process. SU8-2002 was spin coated at 3000 rpm and prebaked at 70 C and 90 C both for 3 minutes each to get 2 micron thickness of SU8.

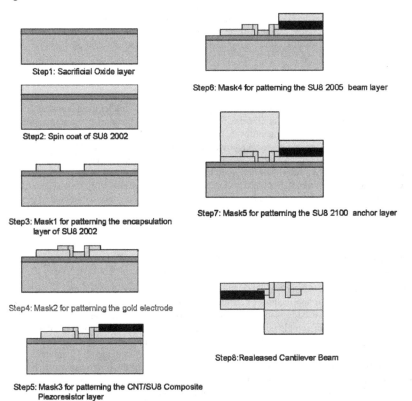

Figure 1. Process flow for Polymer/CNT composite cantilever

The patterned SU8-2002 was acted as an encapsulation layer for SU8/MWNT layer. A post exposure bake was given with same baking parameter like prebaking. The Mask 1 pattern was developed using SU8 developer. In the next step, gold electrode was sputtered and deposited followed by patterning using Mask 2. At this stage, the sample was ready to form the piezoresistive layer of SU8/MWNT. For this purpose MWNT (NC-3100, purified multiwall carbon nanotubes, L/D 100-1000, purity > 95%), procured from Nanocyl CA Belgium, had been modified using octadecyl triphenyl phosphonium bromide (OTPB) as reported earlier [14]. The purpose of modification is to reduce MWNT aggregate size in SU8 matrix and assist in the debundling of MWNT. The debundling leads to an improved dispersion of MWNT in SU8. Sonication, using Misonix Probe Sonicator, was carried out at a power of 4 watts for 20 min. MWNT was initiallysonicated in Microchem nanothinner (cyclopentanone) followed by sonication after adding SU-8 to the MWNT/nanothinner. The result of second sonication was a uniformly dispersed spin coatable SU8 polymer composite with MWNT concentration between 0.05 wt % to 0.2 wt%. After getting a proper dispersion of MWNT in SU8, SU8- 2002 was spin coated and patterned by mask 3 for piezoresistive layer. SU8-2005 was spin coated and patterned using Mask 4 to cover the MWNT/SU8 composite piezoresistor. An anchor layer of thickness of 150 μm using SU8-2100 was coated and patterned using Mask 5. At the final stage of release the sacrificial layer of oxide was etched by buffered hydrofluoric acid and the cantilever die was rinsed in deionized water and isopropyl alcohol. The entire process sequence is shown in figure 1 above. SEM microphotographs of fabricated devices are shown in figure 2. SEM images had been taken at scale of 100μm.

 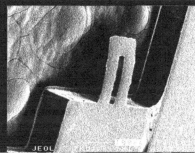

Figure 2. SEM images for fabricated polymer micro-cantilever

RESULTS AND DISCUSSION

The piezoresistors having different OTPB modified MWNT (1:1) concentrations in SU8 were tested without any bending, wherein the neat MWNT concentration was varied from 0.05-0.2 wt%. A programmable voltage source was applied to the piezoresistor and current was measured using a Keithley measurement system. The results are shown in figure 3 and it is observed that reasonable conductivity is obtained only beyond 0.1 wt% of MWNT in MWNT/SU8 piezoresistors.

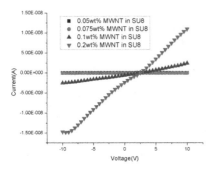

Figure 3. Current versus voltage plot for four different concentration of MWNT in SU8

The piezoresistive action of this fabricated cantilever with 0.1 wt% and 0.2 wt% MWNT in SU8 was measured by deflecting the beam of the cantilever. Initially, the reading of current-vs-voltage had been taken for resistor without bending it. To achieve beam deflection a micrometer screw controlled needle probe was used. The cantilever beam was deflected from 10μm to 50μm with 10μm steps. I-V results for 0.1 wt% and 0.2 wt% MWNT are shown in Fig 4. The change in current is an order of magnitude higher for 0.2 wt% MWNT.

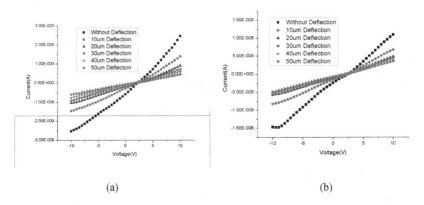

(a)   (b)

Figure 4. Current versus voltage for different deflection of cantilever beam with
(a) 0.1 wt% MWNT in SU8 and (b) 0.2 wt% MWNT in SU8

The relative change of resistance (ΔR/R) with deflection of cantilever beam is shown in Fig 5. The plot shows that the 0.2 wt% MWNT based device is more sensitive than the one with 0.1 wt% MWNT. The gauge factors have been calculated and values are 280 for 0.1 wt% MWNT and 540 for 0.2 wt% MWNT. These are extremely attractive for very sensitive detectors, though linearity has to be improved by performing more experiments with MWNT concentration between 0.1 wt% to 0.3 wt%.

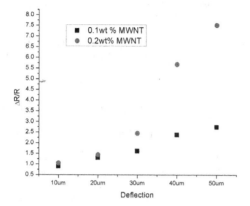

Figure 5. Relative change of resistance as a change in deflection for two different cantilevers with piezoresistor containing 0.1 wt% and 0.2 wt% MWNT in SU8

CONCLUSIONS

The piezoresistor fabricated using MWNT/SU8 composite have shown extremely large gauge factors (upto 280 for 0.1wt% and 540 for 0.2wt% of MWNT in SU8) which are attractive for many bio-sensor applications. However, linearity and sensitivity need to be further optimized by more systematic experiments with MWNT concentrations between 0.1 wt% to 0.3 wt%. Due to very low process temperature, below 90°C, this fabrication process MWNT/SU8 composite based cantilever sensor is compatible with CMOS technologies and it will be possible to realize Systems-on-Chip using a "CMOS first MEMS last" approach.

ACKNOWLEDGMENTS

The authors thank the Centre of Excellence in Nanoelectronics for providing the facilities and support. The Centre of Excellence in Nanoelectronics is funded from Ministry of Communication and Information Technology, Govt of India.

REFERENCES:

1. M. Tortonese, H. Yamada, R. C. Barrett, and C. F. Quate, in *The Proceeding of Transducers '91* IEEE, Piscataway, NJ, 1991, pp. 448-451.
2. J. A. Harley and T. W. Kenny *Appl. Phys. Lett.* 75, 289 (1999)
3. F. J. Giessibl and B. M. *Trafas, Rev. Sci. Instrum.* 65, 1923 (1994).
4. A. Boisen, J. Thaysen, H. Jensenius, and O. Hansen, *Ultramicroscopy* 82,11 (2000)
5. Mo Yang, Xuan Zhang, Kambiz Vafai and Cengiz S Ozkan , *J. Micromech. Microeng.* 13 (2003) pp. 864–872
6. Johnny H. HE and Yong Feng LI, *Journal of Physics: Conference Series* 34 (2006) pp 429–435.
7. Dazhong Jin, Xinxin Li, Jian Liu, Guomin Zuo, YuelinWang, Min Liu and Haitao Yu, *J. Micromech. Microeng.* 16 (2006) pp 1017–1023.
8. J Thaysen, A D Yal cinkaya, P Vettiger and A Menon, *J. Phys. D: Appl. Phys* 35 (2002) pp 2698-2703.
9. N.S..Kale, S.Nag, R.Pinto, V.R.Rao *J. Microelectromech. Syst* 18 (2009) pp79-87
10. Seena.V,N.S.Kale,Sudip.N,M.Joshi,S.Mukherji,V.R.Rao   *Int.J. Micro and Nano Syst.*1(2009) pp.65-70.
11. L. Gammelgaard, P. A. Rasmussen, M. Calleja, P. Vettiger, and A. Boisen, *Appl. Phys. Lett.*, 88 (2006) 113508.
12. Seena V, Anukool Rajorya, Prita Pant , Soumyo Mukherji , V. Ramgopal Rao, *Solid State Sciences* 11 (2009) pp.1606–1611.
13. V.Seena, A. Rajorya, A. Fernaundus, K. Dhale, P. Pant, S. Mukherji, V R.Rao, Proceedings of the 23rd IEEE International Conference on Micro Electro Mechanical Systems (2010), January 24 - 28, 2010, Hong Kong, pp. 851 - 854
14. S. Bose, A.R. Bhattacharyya, R.A. Khare, A.R. Kulkarni, T.U. Patro, and P. Sivaraman, *Nanotechnology,* 19, 335704 (2008).

Mater. Res. Soc. Symp. Proc. Vol. 1299 © 2011 Materials Research Society
DOI: 10.1557/opl.2011.62

Thin film amorphous silicon bulk-mode disk resonators fabricated on glass substrates

A. Gualdino[1], V. Chu[1], and J. P. Conde[1,2]
[1]INESC-MN and Institute of Nanoscience and Nanotechnology, Lisbon, Portugal
[2]Department of Chemical and Biological Engineering, Instituto Superior Técnico, Lisbon, Portugal

ABSTRACT

The fabrication and characterization of thin-film silicon bulk resonators processed on glass substrates is described. The microelectromechanical (MEMS) structures consist of surface micromachined disk resonators of phosphorous-doped hydrogenated amorphous silicon (n$^+$-a-Si:H) deposited by radiofrequency plasma enhanced chemical vapour deposition (RF-PECVD). The devices are driven into resonance by electrostatic actuation and the vibrational displacement is detected optically. Resonance frequencies up to 30 MHz and quality factors in the 10^3-10^4 range in vacuum were measured. A high density of modes that increases with resonator diameter was observed. Membrane-like vibrational modes show good agreement with finite element simulations. The effect of geometrical dimensions of the disks on the resonance frequency was also studied. When operated in air higher harmonic modes show increasing quality factors.

INTRODUCTION

Hydrogenated amorphous and microcrystalline hydrogenated silicon (a-Si:H) thin films have been used in solar cells, thin film transistors in liquid crystal displays, Si based optoelectronics devices and radiation detectors [1]. Thin film a-Si:H has the potential for MEMS applications because of its relatively low stress and low deposition temperature, that can be as low as room temperature (RT) [2,3]. Residual stress can be controlled by tuning the deposition conditions or using stress compensation layers. Low temperature processing allows for the integration of MEMS with electronics as part of the backend processing of CMOS technology. In addition, low temperature processing allows for the use of a wide variety of large area substrates such as glass (transparent, large area and low cost) and plastic (large area, flexible and low cost). The good electronic (with phosphine or boron doped) and mechanical properties properties of a-Si:H film suggest that they can be used as the structural layers for thin film MEMS [3,4].

MEMS resonators are promising alternatives to quartz for timing references [5,6], can be used as radiofrequency (RF) filters [7], and as microbalances in biological and chemical sensor applications [8]. Recent theoretical work using numerical simulations of microscale cantilevers suggest that in-plane modes should yield significantly improved quality factors and sensitivities in air or liquid [9]. Bulk resonators of c-Si and poly-Si show higher quality factors, both in vacuum and in dissipative media, and allow a wider range of tuning of the resonance frequency, than flexural or torsional resonators [10-12]. This work reports on the fabrication and characterization of thin-film hydrogenated amorphous silicon surface micromachined bulk-mode disk resonators processed at temperatures below 250°C on glass substrates.

The movement of the 3D structures can be actuated and detected using several strategies, including optical, capacitive, and piezoelectric or piezoresistive detection. In this work, resonance displacement resulting from electrostatic actuation is optically monitored.

EXPERIMENT

Figure 1 summarizes the fabrication process for bulk resonators with 2 pairs of orthogonally located anchor stems with lateral electrodes. A sacrificial layer of 1 μm thick Al is deposited by DC magnetron sputtering and patterned by wet etching. The structural layer of n^+-a-Si:H is deposited by RF plasma-enhanced chemical vapour deposition (PECVD) with 5 W power and 100 mTorr pressure while flowing a gas mixture of silane, hydrogen and phosphine. After the final metallization with TiW, the structures are patterned by reactive ion etching (RIE), resulting in reasonably vertical sidewalls. In the final step, a wet etchant is used to selectively remove the sacrificial layer and release the structures. Typical dimensions are diameter, D, of 50-300 μm; thickness, h, of 3 μm; and gap distance, g, of 2 μm. Figure 2 shows a scanning electron microscope (SEM) image of a device.

The structures are electrostatically actuated by applying a voltage with DC and AC components between the disk and the electrode. The resulting deflection is detected using an optical method that has been described previously [13]. Distinct electrode geometries were applied for the wineglass and extensional bulk modes, with 1 and 2 pairs of opposed electrodes, respectively. Electrodes are placed around the disk in each of the four quadrants and at the edge of the resonators. To actuate the wine-glass mode, identical signals are applied on opposing electrodes along one axis. For the extensional modes, the signal is also applied to the electrodes on the orthogonal axis. The resonance frequency of the disk resonators, which occurs in the MHz range, is measured at ambient pressure and 10^{-6} Torr, as a function of the diameter.

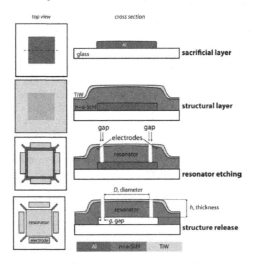

Figure 1. Fabrication sequence of n^+-a-Si:H bulk resonators on glass substrates. A square-shaped resonator was used for simplicity of the illustration.

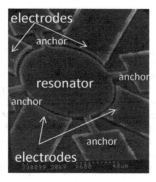

Figure 2. SEM image of a 50 μm diameter disk resonator with 2 pairs of electrodes for extensional mode actuation.

DISCUSSION

Figure 3 shows the frequency response of a 150 μm diameter disk when actuated with $V_{DC} = 30$ V and $V_{AC} = 1$ V, measured at 10^{-6} Torr. A high number of vibrational modes are present, with resonance frequencies from a few kHz up to 26 MHz. The peak intensity is smaller at the higher frequencies, indicating smaller vibrational amplitudes. The quality factor, Q, is defined as $Q = f_{res} / \Delta f_{-3dB}$, where Δf_{-3dB} is the width of the peak 3 dB below its maximum, and $1/Q$ is proportional to the energy dissipated by the microresonator. The resonance peaks have Q values up to ~10 000 (see inset Figure 3).

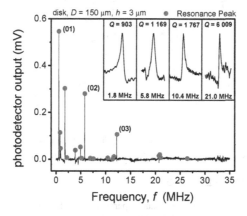

Figure 3. Resonance peaks of a disk 150 μm disk resonator measured at 10^{-6} Torr. Membrane-like modes (01), (02) and (03) are identified ($RBW = 1$ kHz).

FE Modal Analysis

The attribution of the vibrational mode related to the resonance peaks is made using finite-element (FE) mechanical modeling of the MEMS structures. FE simulations enabled the identification of the desired bulk modes, from 10 MHz to 100 MHz, and also flexural membrane-like modes starting at lower frequencies. The resonance peaks (01), (02) and (03) of Figure 3 could be simulated with good agreement and were found to correspond to the fundamental membrane modes with zero, one and two diameters of nodal displacement. Taking a diameter series of microresonators, the flexural nature of these modes is also evident, since the frequency follows $1/D^2$ dependence (see Figure 4). At large disk diameters the effect of residual stresses may be responsible for resonance frequency shifts due to changes in the net structure stiffness. Vibrational modes with combinations of diametric and radial lines of nodal displacement are the solution for other resonant peaks observed. All solutions can be expressed as

$$f_m = \frac{k_m h}{2\pi D^2} \sqrt{\frac{E}{\rho(1-v^2)}} \qquad (1)$$

where E, ρ and v are the Young's modulus, density of the structural material, and its Poisson's ratio, respectively. k_m is a constant determined from the zeros of the Bessel functions that result when solving the standing wave equations for a free membrane [14]. The solutions for in-plane bulk modes result in resonance frequencies that depend on the inverse of the diameter and are independent of the thickness [10]

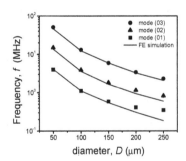

Figure 4. Resonance frequencies of membrane modes (01), (02) and (03) from disk microresonators measured as a function of the resonator diameter, in vacuum.

Energy Dissipation and Quality Factors

The influence of pressure on the resonance parameters of the different oscillation modes observed is shown in Figure 5 (a). The plot shows a single resonance peak, normalized to the resonance amplitude of a disk resonator with 300 μm diameter, measured at ambient pressure and at 10^{-6} Torr. When the pressure is decreased, the maximum amplitude increases significantly, the resonance peak narrows, and there is a small frequency shift. In Figure 5 (b) all the resonance peaks for a single disk resonator are presented, and Q is plotted as a function of the frequency. Under vacuum, the resonance peaks have constant Q values in the range $\sim 10^3$-10^4. Possible limitations of the quality factor are related to dissipation mechanisms such as thermoelastic damping, clamping losses, surface/interface losses, phonon-phonon, and phonon-electron interactions [15-17]. The thermoelastic damping limited quality factor, plotted in Figure 5, is given by [18]

$$\frac{1}{Q_{TED}} = \frac{\alpha^2 TE}{\rho C_p} \frac{2 f \rho C_p t^2 / \pi k}{1+(2 f \rho C_p t^2 / \pi k)^2} \qquad (2)$$

Where T, α, C_p and k are the temperature, linear thermal expansion coefficient, heat capacity and thermal conductivity of the material, respectively. The values reported here are about an order of magnitude below the a-Si:H-TED limit, at the ~ 1 MHz minimum.

When operated in air the Q's drop several orders of magnitude. The damping of oscillating mass due to air viscosity is inversely proportional to the frequency [5] resulting in Q's that start from 10, and increasing linearly, up to ~ 1000. Viscous damping is the dominant mechanism for resonant MEMS in dissipative media. In this work, higher harmonics of resonant modes have increasingly larger quality factors while the fundamental mode has the lowest Q. The inherent differences in Q factors of these modes are likely due to the motion specific nature of the surrounding fluid-microcantilever interactions. Similar behavior has been reported for other microresonators [8,19-20].

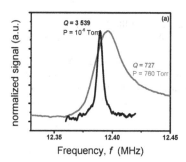 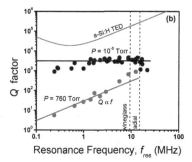

Figure 5. (a) 12,8 MHz resonance peak at 760 Torr and 10^{-6} Torr for a 300 µm diameter disk resonator. (b) Quality factors as a function of resonance frequency, measured in air and vacuum. The two first predicted in-plane bulk modes (wineglass and radial) are identified with the dashed lines.

CONCLUSIONS

This work presents disk microresonators using n^+-a-Si:H thin-films fabricated on glass substrates. The resonance frequencies of these MEMS devices were measured in vacuum and air, with resonance frequencies up to 30 MHz. The microresonators presented in this work have a high surface-to-volume ratio and good potential for high-Q sensors when operated in air. Higher harmonics of the fundamental modes show promise for operation in viscous media because the quality factor increases with frequency. In-plane resonant modes could not be distinguished from membrane like flexural modes since the energy dissipation in vacuum is constant with quality factors in the 10^3-10^4 range.

ACKNOWLEDGMENTS
A. Gualdino thanks Fundação para a Ciência e Tecnologia (FCT) for a doctoral grant (grant no SFRH/BD/48158/2008). This work was funded by FCT through research projects (PTDC/CTM/72772/2006) and the Associated Laboratory IN.

REFERENCES
1. R. W. Collins, A. S. Ferlanto, Current Opinion in Solid State and Mater. Sci. 6, 425 (2000).
2. Y.Q. Fu, J. K. Luo, S. B. Milne, A. J. Flewitt, W. I. Milne, Mater. Sci. Engng. B 124, 132 (2005).
3. P. Alpuim, V. Chu, J. P. Conde, J. Vac. Sci. Technol., 21 (4), 1048 Jul (2003).
4. J. Gaspar, O. Paul, V. Chu, J.P. Conde, J. Micromech. Microeng. 20, 035022 (2010)
5. W. Jing, Z. Ren, C.T.-C. Nguyen IEEE Trans. Ultrason., Ferroelectr., Freq. Control, 51 (12), 1607- 1628 (2004),
6. V. Kaajakari, T. Mattila, A. Oja, J. Kiihamaki, H. Seppa, IEEE Electron Device Lett, 25 (4), 173- 175, (2004).
7. C.T.-C. Nguyen, IEEE Trans. Microw. Theory Tech, 47 (8), 1486-1503, (1999).
8. B. Ilic, H.G. Craighead, S. Krylov, W. Senaratne, C. Ober, P. Neuzil, J. Appl. Phys. 95 (7), 3694-3703 (2004).
9. I. Dufour, S. M. Heinrich, F. Josse, J. Microelectromech. Syst. 16 (1), 44-49 (2007).
10. H. Zhili, S. Pourkamali, F. Ayazi, J. Microelectromech. Syst. 13 (6), 1043-1053 (2004).
11. J. E.-Y. Lee, Y. Zhu, A. A. Seshia, J. Microelectromech. Syst. 18 (6), 064001 (2008).
12. L. Sheng-Shian, L. Yu-Wei; X. Yuan, R. Zeying, C.T.-C. Nguyen, IEEE Ultrasonics Symposium, 3, 1596-1599 (2005).
13. J. Gaspar, V.Chu, and J. P. Conde, Appl. Phys. Lett., 93 (12), 10018-10029 (2003).
14. A. W. Leissa, Vibration of Plates, NASA SP-160 (1969).
15. P. S. Waggoner, C. P. Tan, L. Bellan, H. G. Craighead, J. Appl. Phys., 105 (9), 094315 (2009).
16. A. N. Cleland, *Foundations of Nanomechanics*, (Springer, New York, 2002).
17. J. Yang, T. Ono, M. Esashi, J. Microelectromech. Syst. 11, 775-783 (2002).
18. R. Lifschitz, M. L. Roukes, Phys. Rev. B 61, 5600-5609, (2000).
19. L. B. Sharos, A. Raman, S. Crittenden, and R. Reifenberger, Appl. Phys. Lett. 84, 4638 (2004).
20. S. Dohn, R. Sandberg, W. Svendsen, and A. Boisen, Appl. Phys. Lett. 86, 233501 (2005).

Mater. Res. Soc. Symp. Proc. Vol. 1299 © 2011 Materials Research Society
DOI: 10.1557/opl.2011.56

Fabrication and Characterization of MEMS-Based Structures from a Bio-Inspired, Chemo-Responsive Polymer Nanocomposite

Allison E. Hess[1] and Christian A. Zorman[1,2]

[1] Department of Electrical Engineering and Computer Science, Case Western Reserve University, Cleveland, OH, 44106
[2] Advanced Platform Technology Center of Excellence, Louis Stokes Cleveland VA Medical Center, Cleveland, OH, 44106

ABSTRACT

This paper reports the development of micromachining processes, as well as electrical and mechanical evaluation of a stimuli-responsive, mechanically-dynamic polymer nanocomposite for biomedical microsystems. The nanocomposite, which consists of a cellulose nanofiber network embedded in a poly(vinyl acetate) matrix, was shown to display a switchable stiffness comparable to bulk samples, with a Young's modulus of 3570 MPa in the dry state, which reduced to ~25 MPa in the wet state, with a stiff-to-flexible transition-time dependent on exposed surface area. Upon immersion in phosphate buffered saline, the ac resistance through the PVAc-TW thickness was found to reduce from 8.04 MΩ to ~17 kΩ. Electrochemical impedance of an Au electrode on PVAc-TW was found to be ~178 kΩ at 1 kHz, and this was found to be stable as the probe shank was flexed to compress the metal, but increased with increasing flex angle when the metal was flexed into a tensile state.

INTRODUCTION

Long-term (>20 years) viability of intracortical probes is essential for advancing clinical applications of neural recording technology for improving the quality of life for people with neurological conditions, such as spinal cord injury or Parkinson's disease. However, current probes, which are primarily based on Si,[1,2] are often unable to record neural signals after a few months due to glial scarring that electrically and mechanically isolates electrodes from neurons.[3] It has been hypothesized that strain induced by micromotion between the stiff Si (Young's modulus, E=~160 GPa) and soft cortical tissue (E=~10 kPa) triggers the immune response that produces this cellular sheath,[4] and that a probe based on a material that better matches cortical tissue mechanics would alleviate the issue, allowing for long-term electrode viability. In response, polymer-based intracortical probes using polyimide[5] and parylene[6] have been developed, but have suffered from the inability to penetrate the pia mater without buckling unless stiffened with a rigid backbone or gel-filled microfluidic channel.

A stimuli-responsive polymer nanocomposite, inspired by the switchable stiffness behavior of the sea cucumber (*Cucumaria frondosa*) dermis,[7] has recently been developed by our colleagues for use as a substrate for adaptive intracortical probes for potentially long-term neural recording.[8] Comprised of a soft poly(vinyl acetate) matrix with rod-like cellulose nanofiber fillers (also referred to as whiskers) obtained from filter-feeding sea creatures known as tunicates, the stiffness of the polymer nanocomposite (PVAc-TW) is modulated by interactions between the surface hydroxyl groups on the cellulose nanofibers. In the dehydrated state, hydrogen bonds form between the nanofibers to stiffen the overall material, as measured by a storage modulus of 5.2 GPa in bulk samples; when saturated with hydrogen-bond-forming liquids, the interactions between nanofibers are displaced, reducing the storage modulus to 12

MPa.[9] Thus, in utilizing PVAc-TW as a substrate for an intracortical microprobe for neural recording, a device can be fabricated that is both sufficiently rigid to penetrate cortical tissue, and also provides a better mechanical match to cortical tissue than polyimide- or parylene-based probes.

In this paper, we report a MEMS-based fabrication process for the development of a PVAc-TW-based intracortical probe with switchable stiffness. The electrical and mechanical behavior of PVAc-TW-based microdevices was assessed through benchtop testing.

DEVICE FABRICATION

Intracortical probes from a PVAc-TW substrate, with lithographically-defined metal and capping layer features, were fabricated as shown in the cross-sectional schematics in figure 1. The substrate was prepared (step 1) by first adhering the solution-cast and compressed,[9] 50-100 μm-thick PVAc-TW freestanding film to a bare Si wafer by applying mild pressure to the film on the wafer while heating on a hotplate at 70°C for 3 minutes. Due to the chemical- and temperature-sensitivity of PVAc-TW to acids, bases, organic solvents, and temperatures >120°C, laser-micromachining with a direct-write CO_2 laser was chosen to pattern PVAc-TW. The overall device geometry was patterned into PVAc-TW using a laser power of 0.5 W, a laser speed of 56 mm/s, and a resolution of 1000 pulses per inch. Mechanical test structures did not require any further processing.

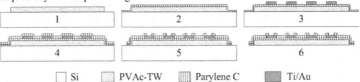

☐ Si   ▨ PVAc-TW   ▥ Parylene C   ■ Ti/Au

Figure 1. Cross-sectional process flow for the fabrication of PVAc-TW-based intracortical probes.

For multilayer structures, a 1 μm-thick parylene interlayer film was vapor-deposited (step 2), coating the top surface and sidewalls of the PVAc-TW microstructures, as well as the exposed areas of the wafer. This parylene film served the dual purposes of protecting the PVAc-TW from subsequent chemical processing, as well as providing a moisture barrier to insulate the metal features from the moisture absorbed by PVAc-TW. Next, a 50 nm-thick Ti adhesion layer and a 200 nm-thick Au layer were sputter-deposited, then patterned through a Shipley 1813 photoresist mask, using Au etchant and a BOE dip (step 3). Acetone was used to remove the photoresist mask, followed by rinsing the wafer with isopropanol. A 1 μm-thick parylene capping layer was deposited (step 4), then patterned with an oxygen plasma through an AZ nLOF 2035 photoresist mask, which also served to pattern the outer geometry of the first parylene layer (step 5). The photoresist was removed by carefully spraying with acetone and isopropanol, followed by drying with an air gun. Finally, the devices were carefully peeled from the wafer (step 6); an example device is shown in figure 2. The Au electrode was exposed through the parylene capping layer at the tip of the 2 mm-long, 180 μm-wide shank, and electrical connection was made through the contact pad on the connector end at the opposite end of the device. When deployed, neural signals would be detected at the electrode, and transmitted to external electronics through a flexible probe packaging system currently in development[10].

Adhesion of the multi-layer, PVAc-TW-based probe was assessed by soaking probes in phosphate buffered saline (PBS) for 60 days at 37°C. Delamination of the layers was not observed during this soak test.

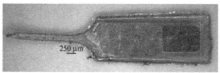

Figure 2. Released PVAc-TW-based probe, with electrode exposed at tip of shank (left) and electrical contact pad on the right.

DIMENSIONAL, ELECTRICAL, AND MECHANICAL RESPONSE TO SWELLING

Swelling-Induced Dimensional Change

Swelling through the PVAc-TW film thickness was compared to swelling laterally across the film by measuring the dimensional change after immersion for 1 hour in DI water at 37°C. It was found that PVAc-TW swelling is highly anisotropic, favoring a through-film dimensional increase by a factor of 12, which is likely a result of the compression step during initial film preparation. This behavior is advantageous because the parylene and metal layers deposited on the PVAc-TW surface do not swell. Isotropic swelling would presumably place the parylene/metal layers under tension, leading to deformation of the device. Since most swelling is through the film thickness, this tensile stress does not arise, and deformation does not occur. Upon saturation in DI water or phosphate buffered saline at 37°C, multilayer structures did not curl, which may be attributed to the anisotropic swelling behavior.

Through-thickness Resistance

The importance of the parylene interlayer was assessed by measuring the ac resistance through the thickness of 2 mm x 2 mm x 50 µm-thick test coupons. Stainless steel wires (75 µm-diam.) were attached with conductive epoxy to 800 µm x 800 µm sputtered-Au bond pads on both PVAc-TW surfaces. The epoxy and bond pads were coated with a 7 µm-thick insulating parylene film. The ac resistance between across the PVAc-TW film was measured with an LCR meter while the sample was immersed in PBS at 37°C. Resistance versus time for a coupon immersed in PBS plot is shown in figure 3. The initial resistance of the dry sample was measured to be 8.04 MΩ, which was quickly reduced upon immersion in PBS. Within ~20 minutes, the resistance was 21 kΩ, and over the next 60 minutes, the resistance stabilized at 17 kΩ. After 100 minutes in PBS, the sample was removed, excess moisture was wicked away with a Kimwipe, and the resistance remained 17 kΩ. These results show that the parylene interlayer is necessary because considerable crosstalk would arise between multiple traces patterned on the PVAc-TW surface during deployment, and exposure of traces and electrodes to the electrolyte absorbed by PVAc-TW would aid in metal corrosion. Thus, for even acute applications, the parylene interlayer between the PVAc-TW and metal electrodes is necessary.

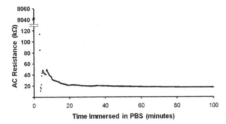

Figure 3. AC resistance measured through 50 μm-thick PVAc-TW film as a function of time exposed to phosphate buffered saline.

Mechanical Characterization

A custom-built microtensile tester was fabricated, using commercial components, to allow measurement of samples with a large elongation limit (>20%) and widely varying Young's modulus[11]. A reservoir was created to fit around tensile samples and keep them wet during testing. Dogbone-shaped samples with beam dimensions 3000 μm-long x 250 μm-wide x 50 μm-thick and grip pads 1500 μm x 1500 μm were tested in one of two modes: "tensile testing" mode, in which the sample was pulled to break or until reaching the 200% strain range of the instrument; and "cyclic testing" mode, in which the sample was alternately loaded and unloaded within the elastic region, allowing for determining the Young's modulus as a function of time.

Tensile testing results, shown in figure 4(left) demonstrate the dramatic difference in Young's modulus in stiff, dry samples (E=3570 MPa) compared to the highly compliant wet samples (E=~25 MPa), which had been immersed in DI water at 20°C for at least one hour prior to testing. This demonstrates that the dynamic behavior translates to microscale samples subjected to micromachining processes.

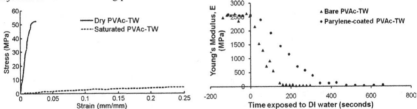

Figure 4. (left) Stress-strain plot of dry (E=3420 MPa) and water-saturated (E=20 MPa) PVAc-TW samples demonstrating dynamic response to water; (right) Young's modulus of bare and parylene-coated PVAc-TW mechanical test structures as a function of time immersed in DI water, showing decreased stiffness as a function of time.

Samples for cyclic testing were placed in the tensile tester while dry, but after approximately five cycles, the reservoir was filled with DI water, completely immersing the sample, while the loading-unloading cycles continued. The Young's modulus was measured on each loading portion of the cycle. Cyclic testing curves for a bare PVAc-TW mechanical sample and a PVAc-TW sample coated with 1 μm parylene are shown in figure 4(right). The bare PVAc-TW sample required approximately 4 minutes of immersion in DI to display its full dynamic range, while the parylene-coated samples required approximately 8 minutes of immersion in DI water to display a comparable range. As the rate of change is dependent upon diffusion into the sample, and because the parylene-coated sample eliminates approximately ½ of the available sample surface area for moisture absorption, it is expected that a parylene-coated sample may display a slowed change in Young's modulus upon exposure to DI water.

ELECTRICAL CHARACTERIZATION

The two-electrode electrochemical impedance spectra of a PVAc-TW-based probe was measured versus a Pt mesh reference electrode in phosphate buffered saline to assess the performance of the electrode for neural recording purposes. The results are shown in figure 5(left). The impedance at 1 kHz, a relevant frequency in neural electrophysiology, was measured to be 178+/- 8 kOhm, which is consistent with Au electrodes of similar size on substrates such as Si,[12] suggesting that the PVAc-TW substrate does not significantly influence the interaction of the Au electrode with the electrolyte.

Because the PVAc-TW substrate was designed to flex and move with the brain after softening, the electrical behavior of a flexed probe was assessed by measuring the impedance of the probe at 1 kHz in PBS while flexing the device about the top of the shank, where it met the connector end. Flexing was controlled by a computer-operated stepper motor, which bent the shank in 14.4 degree increments. Impedance magnitude and phase angle versus flexing angle are shown in figure 5(right), where positive flexing angles correspond to placing the metal under compression, and negative flexing angles correspond to placing the metal under tension. Flexing through positive angles up to 86° did not significantly influence the impedance. However, increased flexing through negative angles corresponded to an increase in impedance, likely a result of stretching the metal trace around the swollen substrate. Thus, probes should be implanted such that bending through positive angles is preferred over bending in the opposite direction.

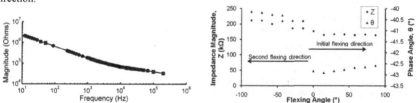

Figure 5. (left) Impedance spectra of a 4500 μm² Au electrode on a PVAc-TW substrate; (right) impedance of Au electrode as a function of flex angle.

CONCLUSIONS

An intracortical probe based on a chemo-responsive polymer nanocomposite with switchable stiffness was fabricated using a combination of laser-micromachining and standard MEMS processing techniques. Measurement of the through-plane ac resistance while immersed in phosphate buffered saline demonstrated that the nanocomposite absorbs the ion-containing electrolytic fluid and becomes conductive, necessitating the use of a thin moisture barrier interlayer between the PVAc-TW substrate and the metal electrodes to prevent trace corrosion and crosstalk between electrodes. Mechanical testing showed that the dynamic behavior was exhibited in the micromachined samples, and that the behavior was also maintained in samples coated with parylene on one surface, though the rate of Young's modulus change was reduced by a factor of two. Finally, electrochemical impedance spectra showed that on a PVAc-TW substrate, Au electrodes behave as they do on Si substrates, and that when flexing the probe to compress the metal, flexing does not have an effect on probe impedance, but the impedance increases when flexing the metal in the tensile state. The fabrication methods presented here can be extended to other moisture-absorbing materials that cannot be exposed to wet chemicals used in lithographic processes. The intracortical probes presented here may be used to evaluate the hypothesis that probe mechanics influence the immune response to a device implanted in the brain.

ACKNOWLEDGMENTS

This work was funded by NSF Grant ECS-0621984, NIH Contract R21-NS053798, and Advanced Platform Technology Center of Excellence of The Department of Veteran's Affairs. The authors acknowledge Prof. S. Rowan and Prof. C. Weder of the Department of Macromolecular Science and Engineering at CWRU, and their research groups, for the nanocomposite material.

REFERENCES

1. K. Najafi, K. D. Wise and Y. Mochizuki, IEEE Transactions on Electron Devices **32**, 1206 (1985).
2. C. T. Nordhausen, E. M. Maynard and R. A. Normann, Brain Research **726**, 129-140 (1996).
3. D. H. Szarowski, M. D. Andersen, S. Retterer, A. J. Spence, M. Isaacson, H. G. Craighead, J. N. Turner and W. Shain, Brain Research **983**, 23-35 (2003).
4. H. Lee, R. V. Bellamkonda, W. Sun and M. E. Levenston, Journal of Neural Engineering **2**, 81 (2005).
5. K.-K. Lee, J. He, A. Singh, S. Massia, G. Ehteshami, B. Kim and G. Raupp, Journal of Micromechanics and Microengineering **14**, 32 (2004).
6. S. Takeuchi, D. Ziegler, Y. Yoshida, K. Mabuchi and T. Suzuki, Lab on a Chip **5**, 519-523 (2005).
7. G. Szulgit and R. Shadwick, J Exp Biol **203**, 1539-1550 (2000).
8. J. R. Capadona, K. Shanmuganathan, D. J. Tyler, S. J. Rowan and C. Weder, Science **319**, 1370-1374 (2008).
9. K. Shanmuganathan, J. R. Capadona, S. J. Rowan and C. Weder, ACS Applied Materials & Interfaces **2**, 165-174 (2009).
10. A. Barnes, A. Hess, C. Zorman, Proceedings of the Technical Conference – 2010 Surface Mount Technology Associated International Conference, 2010.
11. A. Hess, J. Dunning, J. Harris, J. R. Capadona, K. Shanmuganathan, S. J. Rowan, C. Weder, D. J. Tyler and C. A. Zorman, Proceedings of the Solid-State Sensors, Actuators and Microsystems Conference (Transducers) 2009.
12. K. D. Wise and J. B. Angell, Biomedical Engineering, IEEE Transactions on **BME-22**, 212-219 (1975).

Material and Device Reliability

Mater. Res. Soc. Symp. Proc. Vol. 1299 © 2011 Materials Research Society
DOI: 10.1557/opl.2011.251

Characterizing the effect of uniaxial strain on the surface roughness of Si nanowire MEMS-based microstructures.

E. Escobedo-Cousin[1], S.H. Olsen[1], T. Pardoen[2], U. Bhaskar[2], J.-P. Raskin[2]

[1] Newcastle University, Newcastle upon Tyne, United Kingdom
[2] Université Catholique de Louvain, Louvain-la-Neuve, Belgium

ABSTRACT

This work addresses the paucity of roughness measurements by reporting on roughness parameters in uniaxial strained Si beams relevant for state of the art MOSFETs, nanowire and MEMS devices, with varying degrees of strain. Roughness is characterized by high resolution AFM and strain is characterized by Raman spectroscopy. Microstructures comprising a silicon nitride actuator are used to induce a wide range of stress levels in Si beams. The microstructures also allow the comparison of surface evolution in the strain direction (along the Si beam) compared with the unstrained direction (across the Si beam). A gradual reduction in rms roughness amplitude and increase in roughness correlation length in the direction of the applied stress are found for increasing values of strain. In contrast, surface roughness in the direction perpendicular to the applied stress remained largely unchanged from the unstrained initial state.

INTRODUCTION

This work uses an original MEMS concept to strain released silicon beams in order to analyze the relationship between on-chip applied strain and nanoscale surface roughness in Si. Roughness affects carrier mobility through surface roughness scattering at high electric field operation in metal-oxide-semiconductor field-effect transistors (MOSFET). Nanoscale roughness parameters such as rms height (Δ) and correlation length (Λ) are key parameters to model the carrier mobility at high electric fields. Simulations indicate that a reduction in Δ and Λ compared with bulk Si values can explain the high values of electron mobility observed experimentally in tensile strained silicon devices. However, due to the limited characterization techniques available to measure roughness accurately on a nanoscale, to date such assertions have remained largely unconfirmed. Δ can be calculated by atomic force microscopy (AFM) and accurate values of Δ are available to device modellers for several degrees of biaxial strain. In contrast, the determination of Λ requires further work and its actual impact on mobility is not fully understood, therefore, Λ is often used in simulations as an adjustment parameter between theoretical and experimental data.

Two recent works have successfully found experimental correlation between surface roughness parameters and the device performance of biaxially strained Si MOSFETs. Bonno *et al* used high resolution AFM measurements to compare the surface of the strained and unstrained Si, and a statistical analysis of the images was used to extract the roughness parameters [1]. Both Δ and Λ were found to reduce in the presence of strain, although only one level of strain was studied (0.8%). A similar reduction of roughness parameters in strained Si layers was found by Zhao *et al* [2], using high resolution transmission electron microscopy (TEM) to characterize the channel/SiO_2 interface at atomic spatial resolution. The range of strain values studied was between 0.4 and 1.6%.

Although TEM offers a spatial resolution of the surface beyond the conventional capabilities of the AFM, each image only covers an extremely reduced section of the surface. In addition, sample preparation is time consuming, which restricts the sample area that can be studied. In contrast, AFM can cover a larger area on the surface but the measurement resolution is limited by the tip size. However, recent advances in AFM tip technology have enabled the fabrication of ultra sharp tips, making the AFM suitable to the nanoscale characterization required for surface roughness studies on high mobility devices.

To date, surface roughness studies have been carried out on biaxially strained material. The material characterization in uniaxially strained devices is challenging since strain is introduced during device fabrication, therefore, it cannot be investigated separately. As a result, device modeling of uniaxially strained devices has traditionally used roughness parameters extracted from biaxially strained layers.

This work addresses the paucity of roughness measurements by reporting on roughness parameters in uniaxial strained Si beams relevant for state of the art MOSFETs, nanowire and MEMS devices, with varying degrees of strain. MEMS-based microstructures have been used to study the mechanical properties of thin films such as aluminium, polysilicon and silicon nitride [3, 4]. The same microstructures principle is used in this work to induce uniaxial strain in the Si beam by means of a silicon nitride actuator.

EXPERIMENTAL DETAILS

Fabrication of MEMS-based microstructures

Figure 1 shows a schematic diagram of the structure of the MEMS-based microstructures used in this work. The fabrication is detailed elsewhere [3, 4]. Briefly, a silicon-on-insulator (SOI) wafer is used as starting material. The capping Si layer is patterned to generate the Si beam and the buried oxide is used as a sacrificial layer. The process begins with the thermal oxidation and subsequent patterning to outline the Si beam. A second thermal oxidation is carried out and a contact window is opened at the end of the Si beam. Silicon nitride deposition and patterning outlines the actuator beam, which makes contact with the end of the Si beam through the window opened in the second thermal oxide layer. The process finishes with the etching of the oxide box, releasing the microstructure. The thickness of the Si beam and the actuator after release is 130 nm and 225 nm, respectively. The buried oxide had a thickness of 400 nm, which gives the elevation of the beams over the Si substrate.

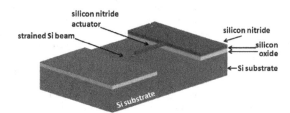

Figure 1. Structure of the MEMS-based microstructure

The strain in the Si beam is induced by the tensor due to its residual stress at high temperature deposition of the silicon nitride. When it cools down to room temperature, the

156

silicon nitride contracts, stretching the Si beam. The strain in the beam was determined by measuring the displacement u of the Si beam/actuator contact region with respect with its original location. Embedded cursors have been positioned on the beams and sidewalls for accurate displacement determination. The strain can be calculated as follows:

$$\varepsilon = \frac{u}{L_0} - \varepsilon_{mis}$$

where L_0 is the initial length of the actuator and ε_{mis} is the initial mismatch developed during deposition process [3]. The resolution limiting factors for the extraction of strain based on the displacement measurements are addressed in [5]. The main limitation is associated with the lithography resolution during the structures fabrication. Incorrect initial cursor positioning and undesired etching at the cursor edges may impact the precise determination of u. Nevertheless, the suggested accuracy of the technique (~ 50 nm) is very small compared with the actual beam displacement, which is in the order of microns, and only represents a maximum error of about 2.5% in the determination of strain. Raman spectroscopy was used to confirm the degree of strain in the Si beams determined by displacement measurements. The range of strain values studied was from 0.17% to 2.77%.

Since the actuator length is about three orders of magnitude greater that its width, it is reasonable to assume that deformation occurs in a linear way. The same assumption is valid for the Si beam since stress is applied in only one direction, therefore, the tensile strain in the beam can be considered to be uniaxial. Although the tensile strain in the Si beam also results in a contraction in the direction transverse to the stretching due to the Poisson's effect (i.e. a slight compressive strain across the beam), the magnitude of the contraction is very small compared with the tensile strain along the beam. In addition, analysis of the AFM images was carried out in the direction along the beam, as explained in the following section.

AFM setup and calibration

AFM characterization was carried out using a XE-150 tool from Park Systems. Areas of 750×750 nm^2 were scanned with a resolution of 2048 pixel/line, which resulted in a step of 3.7 Å between consecutive data points. All measurements were performed in non-contact mode using a super sharp Si tip. This setup ensures large area coverage while maintaining a high scan resolution to allow the detailed study of specific regions. Although noise conditions were minimised in an acoustic isolation chamber, a simple median filter was applied to the image to further suppress any noise contributions. Without median filtering, the noise floor of the equipment is ~ 0.02 nm. Λ was calculated on each image following the correlation analysis described in [1, 6], assuming an exponential surface distribution for the correlation function. Both AFM image acquisition and analysis have been performed in the direction along the Si beams in order to concentrate in the Λ components induced by the tensile strain only.

Since image artifacts may impact the results, only artifact-free regions on each AFM image have been analysed. Figure 2 shows the AFM scan of a Si beam. The left side of the image appears stretched due to thermal drift in the scanner. This effect is more pronounced in small scan areas. The image also exhibits a large surface feature, which could be either a dust particle or a residue from the processing, and several surface pits which were ignored during data analysis since they can cause inaccurate values of Δ and Λ.

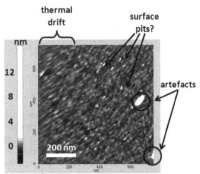

Figure 2. 750 × 750 nm^2 AFM scan of a strained Si beam exhibiting artifacts.

The lateral resolution of the measurements was verified using the surface pits shown in figure 2. Figure 3a shows an amplified image of a region around a surface pit. The corresponding line profiles in figure 3b demonstrate that the AFM tip was able to resolve the width at the bottom of the pit down to ~ 2 nm, in good agreement with the tip radius specification (< 5 nm).

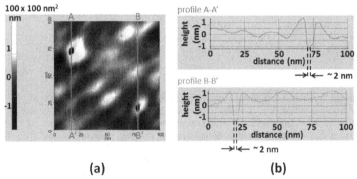

(a)   **(b)**

Figure 3. (a) 100 × 100 nm^2 region highlighting a surface pit in the strained Si beam; (b) line profiles across the pits.

RESULTS

The impact of processing was assessed by comparing the starting blanket SOI material with SOI regions in the sample following processing. Figure 4a compares AFM scans on blanket and processed SOI. The surface of the blanket SOI sample appears smoother than that of the processed material, which also exhibits a high concentration of hillocks likely resulting from the fabrication. In addition to the hillocks, surface roughening is evident over the rest of the surface. Roughness parameters are plotted in figures 4b and 4c, showing considerably higher values of Δ and Λ in the processed material compared with blanket. Nevertheless, process induced roughening should impact all Si beams equally since fabrication was simultaneous. In addition, the surface roughening in the processed SOI is likely introduced by the formation of the thermal oxide, which is also common during MOSFET gate processing and may have a similar impact on device interface roughness.

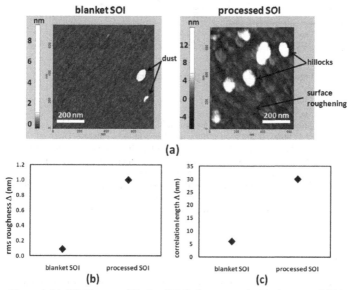

Figure 4. (a) AFM images of blanket SOI (before processing) and processed SOI; (b) the impact of processing on Δ; (c) impact of processing on Λ.

Figure 5 shows the quantification of Δ and Λ on various strained Si beams as a function of strain. Increasing strain values from 0.17% to 2.7% clearly result in decreasing Δ from 0.34 nm to 0.2 nm. In contrast, the evolution of Λ with strain is less clear: for strain values between 0.17% and 1.7%, Λ maintains relatively constant values around 5 nm, however, there is a sudden change at higher strain regimes. Although it is possible that significant topography modifications are only initiated at high strain conditions, more detailed analysis in specific regions of the surface can help to eliminate the any contributions of long scale undulations such as the process induced lumps observed in figure 4a.

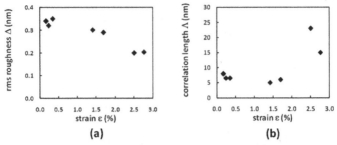

Figure 5. Δ and Λ as a function of strain in the Si beam.

SUMMARY

MEMS-based microstructures comprising a strained Si beam have been used to characterize the impact of uniaxial strain on surface roughness at the nanoscale. Uniaxial strain is induced in the Si beam through the residual stress of a silicon nitride actuator after high temperature deposition. The nanoscale characterization of roughness parameters is key to modeling of state-of-the-art MOSFETs. Results confirm the gradual reduction of rms roughness with increasing strain. Correlation length values are also affected by strain, albeit not in the same gradual trend as the rms values. Further analysis of specific regions may be necessary in order to eliminate the effect of process induced undulations on the quantification of roughness parameters.

ACKNOWLEDGMENTS

This work has been carried out under the EU FP7 NANOSIL framework. The authors are grateful to Ferran Ureña for technical discussions and the Raman spectroscopy work.

REFERENCES

1. O. Bonno, S. Barraud, D. Mariolle and F. Andrieu, Journal of Applied Physics **103** (6), 063715-063719 (2008).
2. Y. Zhao, M. Takenaka and S. Takagi, Electron Device Letters, IEEE **30** (9), 987-989 (2009).
3. A. Boé, A. Safi, M. Coulombier, D. Fabregue, T. Pardoen and J.-P. Raskin, Smart Materials and Structures **18** (11), 115018 (2009).
4. N. André, M. Coulombier, V. De Longueville, D. Fabrègue, T. Gets, S. Gravier, T. Pardoen and J. -P. Raskin, Microelectronics Engineering **84** (11), 2714-2718 (2007).
5. S. Gravier, M. Coulombier, A. Safi, N. André, A. Boé, J.-P. Raskin and T. Pardoen, Journal of Microelectromechanical Systems **18** (3), 555-569 (2009).
6. S. M. Goodnick, D. K. Ferry, C. W. Wilmsen, Z. Liliental, D. Fathy and O. L. Krivanek, Physical Review B **32** (12), 8171 (1985).

Mater. Res. Soc. Symp. Proc. Vol. 1299 © 2011 Materials Research Society
DOI: 10.1557/opl.2011.55

Mechanism of hole inlet closure in shape transformation of hole arrays on Si(001) substrates by hydrogen annealing

Reiko Hiruta[1], Hitoshi Kuribayashi[2], Koichi Sudoh[3] and Ryosuke Shimizu[4]
[1]Fuji Electric Holdings Co., Ltd., 4-18-1, Tsukama, Matsumoto, Nagano 390-0821, Japan
[2]Fuji Electric Systems Co., Ltd., 4-18-1, Tsukama, Matsumoto, Nagano 390-0821, Japan
[3]The Institute of Scientific and Industrial Research, Osaka University, 8-1, Mihogaoka, Ibaraki, Osaka 567-0047, Japan.
[4]Japan Science and Technology Agency, 5, Sanbancho, Chiyoda-ku, Tokyo 102-0075, Japan

ABSTRACT

We investigated the process of the hole inlet closure in surface-diffusion-driven transformation of arrays of high-aspect-ratio holes on Si(001) substrates. The inlet gradually shrinks while keeping the circular shape because of lateral bulging of the inlet surface. We observed complicated top view morphologies reflecting the four-fold symmetry of the Si(001) surface on the inlet surface. Large {111} and {113} facets are formed in the four equivalent azimuths of the [110], while corrugated patterns arise in the four equivalent azimuths of the [100]. Atomic force microscopy observations reveal that the corrugated pattern is composed of three types of small facets, namely, {110} and two {113} in relation of the mirror symmetry. The corrugated pattern formation is due to the geometrical restriction that there is no stable facet between (001) and (010) in the [010] azimuth. The observed morphological evolution is interpreted as surface-diffusion-driven transformation under constraint of the anisotropic surface energy of Si.

INTRODUCTION

Fabrication technique of three-dimensional micro-/nano-structures is crucial for the development of a wide range of three-dimensional devices, such as trench metal-oxide-semiconductor field-effect-transistor (MOSFET) [1], fin-FET [2] and Si-MEMS [3]. When the Si microstructures are annealed at high temperatures in oxygen free ambient, such as in hydrogen gas ambient and in ultrahigh vacuum, they spontaneously change in shape by surface self-diffusion. Recently, such surface-diffusion-driven transformation of Si microstructures by high temperature annealing has been proven to be useful for fabrication processes of three-dimensional microstructures [4-11]. One of the significant applications using such phenomena is formation of silicon-on-nothing (SON) structures, in which a large plate-shaped cavity is formed under a thin Si layer, by annealing of an array of high-aspect-ratio cylindrical holes [4,5,9,12].

The SON structure is formed by surface-diffusion-driven transformation of a hole array via three stages: (1) hole inlet closure, (2) shape change of the voids formed in the substrate, and (3) coalescence of the voids [12]. To precisely control such spontaneous transformation during annealing, detailed understanding of the morphological evolution of hole arrays is required. Previously we reported an investigation on the mechanism of the evolution of voids in the Si substrate during transformation of hole arrays [12]. In this work, we investigate the hole inlet closure, which occurs in the initial stage of the SON structure formation. This stage plays a crucial role in determining the dimensions of the obtained SON structure, i.e., thicknesses of the cavity and top Si layer.

EXPERIMENT

Periodic square arrays of cylindrical holes with the diameter 1.6μm, spacing 1.0μm, and depth 6μm, were fabricated on n-type CZ-Si (100) substrates (2Ω· cm) by anisotropic reactive ion etching (RIE) with an etching mask of silicon dioxide. The substrates were annealed in 10~60 Torr hydrogen gas ambient at 1100~1150 °C. The structures of the samples were evaluated by scanning electron microscopy (SEM), and atomic force microscopy (AFM).

RESULTS AND DISCUSSIONS

Figure 1 show the morphological evolution of the hole array by annealing. The fabricated hole array before annealing is shown in Fig. 1(a). The structures after annealing at 1150 °C for 30 min are shown in Figs. 1(b) and 1(c). The difference in structure between these two samples is due to the difference in H_2 pressure in the annealing ambient. It is known that the transformation rate increases with decreasing H_2 pressure [5-7]. For the sample annealed in 40 Torr H_2, openings have become fairly narrow [Fig. 1(b)]. For the sample annealed in 10 Torr the inlets of most of the holes are completely closed [Fig. 1(c)]. Voids are buried in the substrate, and a periodic corrugation originated from the initial hole pattern remains on the surface.

A high resolution SEM image shown in Fig. 2 reveals the mesoscopic morphology during closure of the inlet. This sample was annealed at 1150 °C for 10 min in 60 Torr H_2. The sample is viewed along the direction inclined by 20° from the substrate normal towards the [110] direction. The hole inlet shows a complicated pattern reflecting the four-fold symmetry of the Si(001) surface. We find that the morphology of the hole inlet is constructed of large facets and corrugated regions.

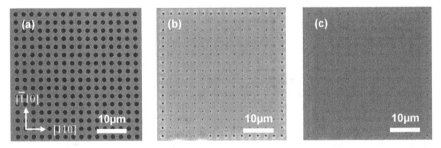

Figure 1. Top view SEM images of the square array of high-aspect-ratio holes fabricated on the Si(001) substrate: (a) as-etched sample, (b) after annealing at 1150 °C for 30 min in 40 Torr H_2, and (c) after annealing at 1150 °C for 30 min in 10 Torr H_2.

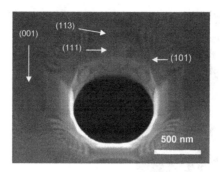

Figure 2. SEM image of a hole after annealing at 1150 °C for 10 min in 60 Torr H_2. The viewing direction is in the direction inclined by 20° from the [001] to [110].

The plane indices of the individual facets seen in Fig. 2 are identified by geometrical analysis based on the thermodynamically stabe facets of Si, namely {001}, {111}, {110}, and {113}. The stereographic projection of these planes is shown in Fig. 3. We determined the plane indices from the arrangement of the facets in the [110] and [010] azimuths. We find that the morphology in Fig. 2 is basically composed of these stable facets. The facets appearing in the [110] azimuth are identified to be (113) and (111) facets as indicated in Fig. 2. In the [010] azimuth, a (011) facet exists with the corrugated pattern. The formation of the corrugated pattern is reasonable, because there is no stable facet between (001) and (011) in this azimuth as found from Fig. 3.

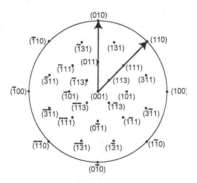

Figure 3. Stereographic projection of {001}, {111}, {110} and {113} planes of Si.

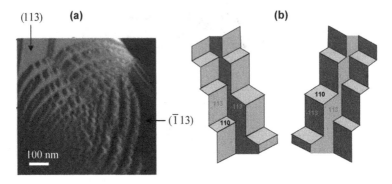

(113) **(a)**   **(b)**

(1̄13)

100 nm

Figure 4. (a) An error mode AFM image of the corrugated pattern formed around the hole inlet. (b) Schematic illustration of the structure of the corrugated pattern.

We use AFM to investigate the structure of the corrugated pattern appearing in the four equivalent azimuthal directions of the [010]. Figure 4(a) shows an error mode AFM image of a corrugated region. In this image the contrast is due to the difference in the local surface orientation. The corrugated pattern is constructed by three types of facets. Allowing for the facet arrangement in the [010] azimuth shown in Fig. 3, we find that the components in the corrugation are (011), (113), and (-113) facets as schematically shown in Fig. 4(b).

Figure 5 shows the evolution of the mesoscopic morphology during inlet closure. The inlet shrinks while keeping the circular shape. The sizes and shapes of {113} facets and the corrugated patterns are rather preserved during the process. The distinct change seen in these SEM images that {111} facets grow around the opening. SEM images in Fig. 6 show the evolution of the cross-sectional morphology normal to the [110] direction. We find that the inlet surface bulges laterally, while the top (001) facets around the inlet are depressed [Fig. 6(a)]. The apex of the bulge is curved rather than faceted, which indicates that the growth rate of the {110} and {010} facets on the apex of the bulge is larger than those of their neighboring facets. We find that the lateral bulging of the inlet surface is caused by the downhill atomic current from the top (001) facets to the apex of the bulge.

CONCLUSIONS

We investigated the morphological evolution during the inlet closure process in the SON structure formation by surface-diffusion-driven transformation of a hole array on Si(001). The top view morphologies during the hole inlet closure show complicated features reflecting the four-fold symmetry of Si(001) surface. The morphologies are composed of large facets and corrugated regions. The morphologies formed during the inlet closure are largely understood allowing for the geometrical restriction due to the anisotropy of the Si surface.

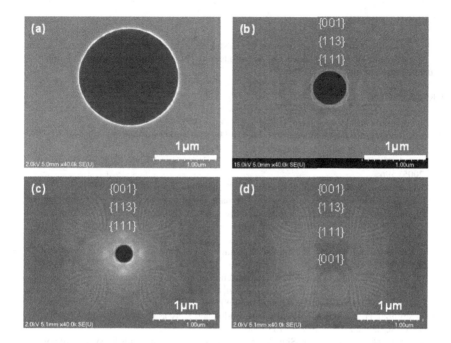

Figure 5. Top view SEM images showing the process of the inlet closure. (a) Structure of the as-etched sample. (b)-(d) Samples after annealing at 1150 °C for 30 min in H_2 pressure of (b) 60, and (c), (d) 40 Torr.

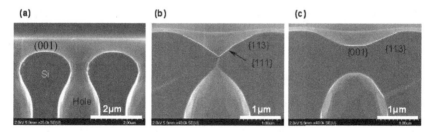

Figure 6. Cross sectional SEM images showing the process of the inlet closure. The viewing direction is [110].

ACKNOWLEDGMENTS

The authors thank Prof. H. Iwasaki of Osaka University for many useful discussions, and Dr. Y. Nagayasu, Dr. M. Nishizawa, Dr. A. Saito and Mr. N. Nakajima for encouraging the implementation of this work.

REFERENCES

1. N. Fujishima and T. A. Salama: IEDM Tech. Dig. (1997) 359.
2. Y. K. Choi, L. Chang, P. Rande, J. S. Lee, D. Ha, S. Balasubramanian, A. Agarwal, M. Ameen, T. J. King and J. Borkor: IEDM Tech. Dig. (2002) 259.
3. M.C. Lee and M.C. Wu: Proc. 18th IEEE Int. Conf. on Micro Electro Mechanical Systems (2005) 596.
4. I. Mizushima, T. Sato, S. Taniguchi, and Y. Tsunashima, Appl. Phys. Lett. 77, 3290 (2000).
5. T. Sato, I. Mizushima, S. Taniguchi, K. Takenaka, S. Shimonishi, H. Hayashi, M. Hatano, K. Sugihara and Y. Tsunashima, Jpn. J. Appl. Phys. 43, 12 (2004).
6. H. Kuribayashi, R. Shimizu, K. Sudoh, and H. Iwasaki, J. Vac. Sci. & Technol. A22, 1406 (2004).
7. H. Kuribayashi, R. Hiruta, R. Shimizu, K. Sudoh and H. Iwasaki: J. Vac. Sci. Technol. A 21 (2003) 1279.
8. M. M. Lee and M. C. Wu, J. Microelectromech. Syst. 15, 338 (2006).
9. V. Depauw, O. Richard, H. Bender, I. Gordon, G. Beaucarne, J. Poortmans, R. Mertens, and J. –P. Celis, Thin Solid Films 516, 6934 (2008).
10. R. Shimizu, H. Kuribayashi, R. Hiruta, K. Sudoh and H. Iwasaki, Proc. 2006 Int. Symp. Power Semicon. Dev. & ICs, p.113 (2006).
11. R. Hiruta, H. Kuribayashi, R. Shimizu, K. Sudoh and H. Iwasaki: M.R.S. Symp. Proc. 958 (2007).
12. K. Sudoh, H. Iwasaki, R. Hiruta, H. Kuribayashi, and R. Shimizu, J. Appl. Phys. 105, 083536 (2009).

Mater. Res. Soc. Symp. Proc. Vol. 1299 © 2011 Materials Research Society
DOI: 10.1557/opl.2011.466

Characterization of Hydrophobic Forces for in Liquid Self-Assembly of Micron-Sized Functional Building Blocks

M. R. Gullo[1], L. Jacot-Descombes[1], L. Aeschimann[2], J. Brugger[1]

[1]Microsystems Laboratory, Institute of Microtechnology, Ecole Polytechnique Federale de Lausanne (EPFL), 1015 Lausanne, Switzerland
[2]Nanoworld AG, 2000 Neuchâtel, Switzerland

ABSTRACT

This paper presents the experimental and numerical study of hydrophobic interaction forces at nanometer scale in the scope of engineering micron-sized building blocks for self-assembly in liquid. The hydrophobic force distance relation of carbon, Teflon and dodeca-thiols immersed in degassed and deionized water has been measured by atomic force microscopy. Carbon and dodeca-thiols showed comparable attractive and binding forces in the rage of pN/nm^2. Teflon showed the weakest binding and no attractive force. Molecular dynamic simulations were performed to correlate the molecular arrangement of water molecules and the hydrophobic interactions measured by atomic force microscopy. The simulations showed a depletion zone of 2Å followed by a layered region of 8Å in the axis perpendicular to the hydrophobic surface.

INTRODUCTION

Capillary forces have often been used to perform templated self-assembly (SA) [1] and folding [2] of micron-sized building blocks. However capillarity has proven to be hardly controllable and thus not fitted to achieve controlled and selective SA. This is even more the case for the SA of a defined number of functional building blocks inside the bulk of the liquid. An alternative and more controlled way to self-assemble micron-sized building blocks would be to use the hydrophobic force. It has been proven that specific and orthogonal hydrophobic patterning can be applied to target areas of the building blocks by local chemical functionalization [3]. Moreover hydrophobic gradients might be engineered to guide the SA and induce the self-repairing of miss aligned assemblies [4]. In order to optimize the design of such functionalized building blocks it is necessary to understand the nature of the hydrophobic force. Especially the force to distance relation of the attraction and the magnitude of binding force is of great interest [5, 6]. The surface force apparatus has extensively been used to study hydrophobic interactions. However it is limited in the materials that can be measured. One way to overcome this limitation is to use atomic force microscopes (AFM) to perform the force distance measurements. AFM is the first choice for many biologists to precisely measure protein unfolding forces and has proven to be a very reliable and sensitive tool [7]. The force distance measurements in this paper are exclusively done by AFM.

There are many models that aim to explain the hydrophobic interaction: a) entropic effects due to molecular rearrangement of water, b) electrostatic effects, c) submicroscopic bubbles and d) cavitation [5, 6, 8]. Each one of these models describes only a part of the highly complex nature of the hydrophobic interaction. The final model will probably be a smart combination of the above mentioned theories [5]. This paper will mainly focus on the entropic model (a). This model can be used to investigate the very short range component of the hydrophobic interaction for degassed and deionized water [5]. A corresponding numerical approach using molecular dynamic (MD) simulations has been developed.

EXPERIMENTAL DETAILS

Dedicated AFM probes with a spherical tip featuring a radius of 2μm (SphereTips, Nanoworld AG) and gold coated tips with 15nm radius (QuartzTip, Nanoworld AG) have been used to perform the force distance measurements (see figure 2d and figure 3b). Silicon chips diced in 1x1cm^2 squares were used as samples. Samples and AFM probes were cleaned in a sequence of filtered acetone, isopropanol alcohol and deionized (DI) water followed by O_2 plasma. Both were simultaneously coated with hydrophobic layers immediately after cleaning. Tree different coatings have been applied: 4nm Evaporated teflon (C_4F_8); 5nm sputtered carbon; dodeca-thiols (Sigma Aldrich) monolayer absorbed on a previously evaporated 50nm thin Au layer. The force distance curves have been acquired with a Nanoscope III picoforce AFM (Veeco) operated in static mode. The liquid medium was filtered and degassed DI water. Before each experiment the force constant of each AFM probe was measured and calibrated by the thermal noise method [9]. The AFM probe approached the sample at a speed of 0.5Hz and was held on the surface for 0.5s before being retracted. In order to acquire a statistically representative amount of curves the sample was approached at 100 different positions and the experiment was redone with several tips and samples.

MD simulations were performed using the LAMMPS software package (Sandia National Laboratories). So far only the interaction of carbon as hydrophobic layer has been modeled. The water molecules were designed based on the SPC/E model [10]. The water carbon interaction was modeled by the optimized potentials for liquid simulations (OPLS) model [11, 12]. A boro/thermostat couple held the system at ambient conditions (1 bar, 293°K). First a calibration experiment was performed. A 3x3nm^2 square slab of water molecules was put on a 10x10nm^2 substrate of carbon atoms (see figure 1). The system was allowed to relax for 500ps in order to form a stable droplet on the substrate. The resulting contact angle was measured and compared to the measured contact angle in order to calibrate the system. In a second experiment the same square lattice of water was held between two layers of carbon. The system was again allowed to relax for 500ps before approaching the surfaces towards each other.

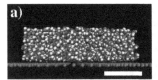 

Figure 1. a) The initial setup of the MD simulation system is a 3x3nm^2 square slab of water molecules (white and blue) on a 10x10nm^2 substrate of carbon atoms (red); the scale bare is 1nm wide. b) The same system after 500ps relaxation time shows the formation of a droplet with a typical hydrophobic contact angle); the scale bare is 1nm wide.

DISCUSSION

The histograms and fits for the hydrophobic attractive and binding force for the carbon coated tip and sample are shown in figure 2a) and b) respectively. The histograms were generated from 500 force curves taken on 100 different positions across the sample and repeated with 3 different sample-tip pairs. The histograms clearly show a Gaussian distribution with a peek at 3.69nN and 53.5nN for the attractive and binding force respectively. The error of about 25% is relatively important. This could be due to the non-uniformity of the sample surface, which could lead to different contact areas between the tip and sample. The approach curves for the teflon coated tip and sample didn't show any attractive force. Figure 2c) shows the histogram and fit of the hydrophobic binding force for the Teflon coatings with a peek at 13.37nN. Compared to the carbon coating the binding force is lower. However the relative error is slightly smaller. This could be due to the softness of the Teflon coating. Therefore the tip indents the layer that might results in a more uniform contact area.

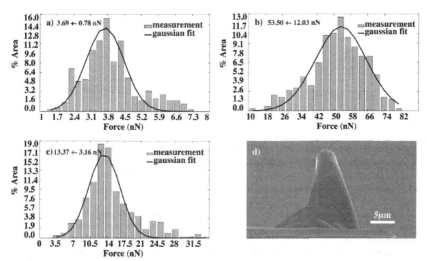

Figure 2. a) Histogram of attractive forces for the carbon coated sample and tip. The Gaussian fit gives a mean value of 3.69 nN with an error of 0.78nN. b) Histogram of binding forces for the carbon coated sample and tip. The Gaussian fit gives a mean value of 53.5 nN with an error of 12.03nN. c) Histogram of attractive forces for the teflon coated sample and tip. The Gaussian fit gives a mean value of 13.37 nN with an error of 3.16 nN. d) Scanning electron micrograph of the spherical AFM tip used for the experiment (SphereTips, Nanoworld AG). The tip has a radius of curvature of 2μm.

The measurements with the dodeca-thiols coatings were performed with a sharp tip (figure 3b) in order to minimize the effect of the surface roughness resulting from the evaporated Au layer. While the attractive force was constant over time (0.1 ± 0.05 nN) the binding force diminished over time. Figure 3a shows a typical time dependence of the hydrophobic binding in function of the recorded force distance curve. When taking into account the approach speed and resting time mentioned before the time between each curve is 3s. The magnitude of the binding force decreased with each approach. After the system was left to rest for 1h the magnitude of binding force increased again, and followed the same decrease again with each consecutive measurement. The drop in force can be explained by the rearrangement of the thiols previously displaced from the contact area due to the contact interaction force during the AFM experiment. The difference between the two peak values might be attributed to the initial calibration procedure of the AFM, which requires surface contact.

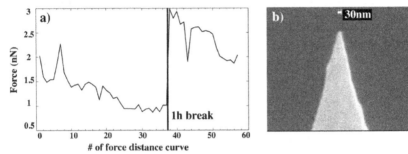

Figure 3. a) Measured hydrophobic binding force for the dodeca-thiols coating in function of the number of the approach curve. A decrease in magnitude can be observed. The line indicates the point at which the system was left to rest for 1h. The consecutive measurements showed an initial increase in force with subsequent decrease. b) Scanning electron micrograph of the AFM tip used for the experiment (QuartzTip, Nanoworld AG). After the gold coating the tip has a radius of curvature of 15nm.

Table I shows the resumed hydrophobic attractive and binding forces for the three coatings normalized to the respective contact areas. The contact surface was calculated for an estimated indentation depth of 2nm, 4nm and 10nm for the carbon, Teflon and dodeca-thiols coating respectively. The force values of the dodeca-thiols are the peak values. The carbon and the dodeca thiols show comparable forces. This could also be explained by the similarity of the methyl end groups of the thiols and the carbon atoms hydrated in DI water.

Table I. Attractive and binding force for the three coatings normalized to the contact area.

Material	Attraction Force	Binding Force
Carbon	0.15 ± 0.03 pN/nm^2	2.12 ± 0.45 pN/nm^2
Teflon (C_4F_8)	None	0.27 ± 0.06 pN/nm^2
Dodecathiols	0.1 ± 0.05 pN/nm^2	2.98 ± 0.19 pN/nm^2

Figure 4 shows the simulated distribution of oxygen and hydrogen atoms in a normalized histogram. The distribution is in volume slices along the z-axis perpendicular to the sample surface. It can be nicely seen that there is a depletion zone of about 2Å from the sample surface situated at 2Å and 30Å for the bottom and top layer respectively. Moreover it can be observed that there are mainly oxygen atoms near both hydrophobic surfaces. The distance between the first peaks of oxygen (a) and hydrogen (b) is comparable to the "height" of the water molecule (see figure 4). This suggests that the water molecules in this first layer are closely packed with the two hydrogen atoms pointing away from the carbon surface (see figure 4). This is in accordance with the energetically more convenient configuration for the water molecule in which it forms two hydrogen bonds with other water molecules instead of orienting the dangling bonds towards the non-polar carbon surface. The second oxygen and hydrogen peeks (c) and (d) are spaced by 1.1Å, indicating a less dense layering as depicted in figure 4. The same layering can be observed symmetrically at the top and bottom borders. The non-uniform distribution has a width of roughly 10Å on both sides. Within the bulk the water molecules are distributed uniformly.

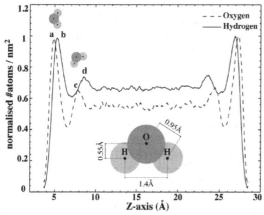

Figure 4. Normalized histogram of the local distribution of hydrogen atoms in volume slices along the z-axis (perpendicular to the sample surface) of the water between the two carbon surfaces. The peaks **a** (first oxygen) and **b** (first hydrogen) are separated by 0.57Å the peaks **c** (second oxygen) and **d** (second hydrogen) are further inside and are separated by 1.1Å. The inset shows the typical distances of the water molecule.

CONCLUSIONS

This work clearly shows that it is possible to quantitatively measure the hydrophobic attraction and binding force for several hydrophobic materials with state of the art AFM techniques. Layers with methyl or carbon end groups showed similar forces. Whereas Fluor terminated layers such as Teflon showed no measurable attraction force and a lower binding force. All forces were in the pN/nm^2 range. In view of controlling the yield and speed of the SA of micron-sized building blocks it is important to be able to influence the hydrophobic interaction. Therefore the dependence of the hydrophobic attractive and binding forces on pH and ionic strength is currently under investigation. MD simulation showed a depletion zone of 2Å and layering of water molecules with different compactness within 10Å from the hydrophobic sample surfaces. Current efforts are made to simulate the approaching of the two hydrophobic surfaces towards each other and then retracting from each other. This will create an interference of the layering zones that could be related to the hydrophobic attractive short-range force and the cavitational hydrophobic binding force [5, 8].

ACKNOWLEDGMENTS

The authors would like to thank Prof. G. Fantner and Dr. B. Erickson from the Laboratory for Bio- and Nano-Instrumentation at EPFL Switzerland for letting us use their AFMs and providing the software to evaluate the force distance curves. We would also like to express our gratitude to Prof. M. Rovere and Prof. W. Andreoni from European Centre of Atomic and Molecular Computation, for their support in all aspects molecular dynamics simulations. We gratefully acknowledge the financial support from the Swiss Confederation and the funding by Nano-Tera.ch within the project SELFSYS.

REFERENCES

1.   M. Mastrangeli, et al., *Journal of Micromechanics and Microengineering* **19**, (2009).
2.   A. Azam, et al., *Biomedical Microdevices* 1 (2010).
3.   T.D. Clark, et al., *Journal of the American Chemical Society* **123**, 7677 (2001).
4.   X. Xiaorong, L. Sheng-Hsiung, and K.F. Bohringer in *Geometric binding site design for surface-tension driven self-assembly*, (Proceedings. ICRA '04. 2004 IEEE International Conference on Robotics and Automation, 2004.) pp. 1141
5.   E.E. Meyer, K.J. Rosenberg, and J. Israelachvili, *Proceedings of the National Academy of Sciences of the United States of America* **103**, 15739 (2006).
6.   D. Chandler, *Nature* **437**, 640 (2005).
7.   T.E. Fisher, et al., *Trends in Biochemical Sciences* **24**, 379 (1999).
8.   K.Q. Fa, A.V. Nguyen, and J.D. Miller, *Journal of Physical Chemistry B* **109**, 13112 (2005).
9.   R. Levy and M. Maaloum, *Nanotechnology* **13**, 33 (2002).
10.   H.J.C. Berendsen, J.R. Grigera, and T.P. Straatsma, *Journal of Physical Chemistry* **91**, 6269 (1987).
11.   T. Koishi, et al., *J. Chem. Phys.* **123**, (2005).
12.   W.L. Jorgensen and J. Gao, *Journal of the American Chemical Society* **110**, 4212 (1988).

Mater. Res. Soc. Symp. Proc. Vol. 1299 © 2011 Materials Research Society
DOI: 10.1557/opl.2011.64

Nanoindentation Characterization of PECVD Silicon Nitride on Silicon Subjected to Mechanical Fatigue Loading

Z-K Huang, K-S Ou, and K-S Chen
Department of Mechanical Engineering National Cheng-Kung University Tainan, Taiwan, 70101

ABSTRACT

In this work, the mechanical properties of PECVD silicon nitride deposited on silicon substrates by two different processing conditions were investigated. Indentation method was primary used for qualitatively examining the effect of process conditions to the achieved mechanical properties. The experimental results indicated that the residual stress, fracture toughness and interfacial strength, as well as the fatigue crack propagation were strongly depended on the processing conditions such as deposition temperatures and chamber pressures. Preliminary results indicated that the specimen deposited at a lower temperature and a lower pressure exhibited a much less residual tensile stress and a better interface strength. On the other hand, it was found that RTA could enhance the interfacial strength but the generated high tensile strength could actually reduce the equivalent toughness and leads to structural reliability concerns. In summary, the characterization results should be possible to provide useful information for correlating the mechanical reliability with the processing parameters for future structural design optimization and for improving the structural integrity of PECVD silicon nitride films for MEMS and IC fabrication.

INTRODUCTION

PECVD nitride films are important and common structural materials used in microsystems applications such as transduction structures, barriers, and mask layers. The mechanical properties of silicon nitride are therefore essential for overall device reliability assessments. In particular, the fatigue and interfacial properties of PECVD nitride coated silicon structures subjected to thermo-mechanical loading will influence the structural longevity of integrated circuits or MEMS actuators. As a result, it is desired to perform characterization to understand the influence of processing and operating conditions such as deposition parameters and thermal annealing as well as service loading for structural longevity concerns.

Our previous investigation [1] on PECVD silicon nitride using nanoindentation technique indicated that the residual stress and the fracture toughness, as well as the interfacial strength were major controlling factors for governing the mechanical reliability issue and these factors strongly depend on deposition and post-deposition thermal processing parameters. In this work, a further investigation was conducted by using PECVD nitrides fabricated by two different process conditions followed by a rapid thermal annealing (RTA). In addition to the residual stress and interface delamination characterizations preformed in our previous work, the mechanical fatigue tests also implemented in this work to further evaluate the longevity of silicon nitride films subjected to repeat mechanical loadings, which could be a concern for MEMS sensors and actuators [1]. Based on the characterization results, it is possible to provide a qualitative conclusion for correlating the mechanical reliability with the processing parameters for leading the future structural design optimization in MEMS devices.

FABRICATION AND EXPERIMENTAL DESIGN

The overall characterization flow is shown in Figure 1. 5000 Å PECVD nitrides were deposited on 4" silicon wafers using a Nano-Architect Research/BR-2000LL system in National Chiao-Tung University. Two different processing conditions were used: the first batch (Specimen A) was deposited at 350°C under a pressure of 5 torr. While the conditions for the other batch (Specimen B) were 300°C with a pressure of 1 torr. The deposition rates were in order of 500 Å per minute. After deposition, the wafers were then experienced a rapid thermal annealing (RTA) between 400 and 800°C with an annealing period 1 minute. Residual stress measurements were then performed by using a KLA-Tencore FLX 2320 curvature measurement system and the data were converted via Stoney's formula [2] to evaluate the influence of RTA on these wafers.

These wafers were then sliced to rectangular shape (35 mm x 10 mm) ready for testing. Several Vickers or Berkovich indentations were indented on the specimen to investigate the possible cracking and delamination for evaluating the fracture toughness and interfacial strength (shown in Figure 2a). Finally, these indented specimens were mounted on a self-designed fatigue system (shown in Figure 2b) to evaluate the possible fatigue crack propagations during service.

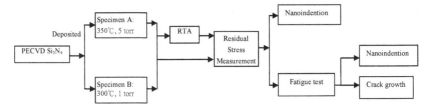

Figue 1 The overall characterization plan

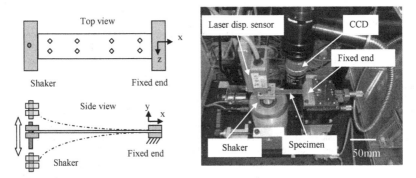

Figue 2 (a) schematic specimen for indentation and fatigue study and (b) the fatigue testing system

EXPERIMENTAL RESULTS
Residual Stress

The residual stress characterization results are shown in Figure 3. The as-deposited residual tensile stresses are 900 MPa and 500 MPa for specimens A and B, respectively. On the other hand, the RTA process at medium temperature (up to 400 - 600°C) tends to increase the residual stresses. However, the residual stresses decrease for further increase in RTA temperatures. For both specimens, the strong difference in the as-deposited residual stresses reveals that the processing parameters have strong impact on the forming and stacking of nitride molecules, which directly related to the residual stress level. Furthermore, the general correlation between the residual stress and the RTA temperature agrees with that reported before [1, 3]. The increase in residual stress is believed to be caused by eliminating volumetric defects and grain boundary and the final stress reduction is due to stress relaxation. However, it is interested to point out that the peak stress temperatures are different for specimens A and B, which indicated that the activation energies for relaxation are different. A more systematic investigation for varying various RTA parameters such as temperature and annealing period for completely characterization is currently underway.

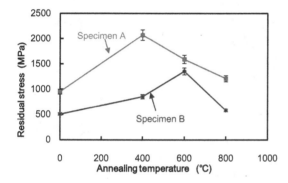

Figue 3 Residual stress characterization results

Nanoindentation Characterization

Although nanoindentation testing has been widely used for characterizing the hardness and Young's modulus of thin films [4], the main objective is to study the fracture properties in this work. Both specimens A and B were nano-indented under a force control manner with a peak load of 300mN and the essential results are shown in Figure 4. The indentation depths were within 15% of film thickness to avoid excessive substrate effect. From Figure 4a, based on Oliver and Pharr's work [4], the hardness and modulus of both specimens are calculated as 89 GPa and 8.7GPa for specimen A, 247GPa and 12GPa for specimen B, respectively. Although these numbers may be deviated from actual value slightly due to substrate effects, it is still possible to observe the significant difference on both physical properties due to different process design based on the data. In addition, the results were also qualitatively agreed with that reported by Huang et al in PECVD nitride deposition research [5], where a higher plasma power or low deposition pressure would result in higher elastic modulus and hardness were reported.

In addition, the pop-out observed in Figure 4a also suggests that the possible delamination occurred during indentation and this can be seen from the SEM micrograph shown in Figure 4b. On the other hand, no pop-out observed for specimen B and the indentation shows no evidence of delamination between the nitride and the substrate. However, the adhesion of specimen A can be improved by applying a one minute 800°C RTA for specimen A. This can also be observed from both the p-h curve and SEM micrograph shown in Figure 4. As a result, an appropriate RTA could effectively enhance the interfacial strength. In addition, by measuring the generated crack length and using the model proposed by Marshall and Lawn [6] and Zhang et al. [7], it is possible to estimate the fracture toughness. It was found that the toughness can also be enhanced by RTA. However, the associated increase in residual tensile actually reduce the capability to resist fracture (i.e., the equivalent fracture toughness)[3].

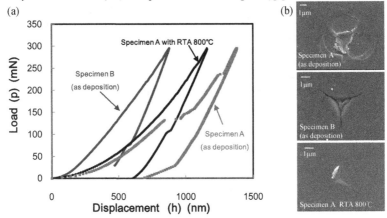

Figue 4 (a) The nanoindentation responses of nitrides with different process parameters and (b) the SEM micrograph of the indentations.

Fatigue Crack Growth

The fatigue crack growth was tested by mounting the pre-indented specimens (crack length a_0) on the testing system. The system then provided a 40Hz sinusoidal fluctuation with various amplitudes. By elementary beam theory and subsequent finite element analyses, the nominal bending stresses at various indentation locations schematically shown in Figure 2a are between 13 and 54 MPa. Using an optical microscopy, it was also possible to measure the propagated crack length a after a certain amount of cycles. Our ultimate goal is to find the fatigue crack growth parameters based on Paris Law [8]. Due to the huge amount of experimental effort required, the complete fatigue characterization is still underway. Nevertheless, it is still possible to make qualitative comparisons between the fatigue behaviors of the two specimens to highlight the influence of process parameters. Figure 5 shows typical experimental data here to briefly demonstrate the observed results. For ease of data representation, the crack growth is normalized by the initial crack length (i.e., $C_p = (a-a_0/a_0) \times 100\%$). In general, a larger remote bending stress would result in a larger crack growth. In addition, it can be seen that specimens A and B exhibit considerable difference behavior in fatigue crack growth. First, for specimen A, crack growth was slower but constantly even the fatigue cycle reached 2 millions. On the other hand, for specimen B, the fatigue crack growth was much smaller even with a higher stress level. However, after treated by a 800°C RTA, the fatigue crack growth of specimen B became much faster and toward saturation for fatigue cycles over 0.5 million cycles.

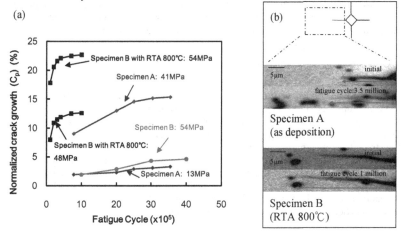

Crack length of specimen A: initially 19μm; after fatigue cycle 3.5million: 22μm
Crack length of specimen B (after RTA): initially 26μm; after fatigue cycle 1million: 33μm

Figue 5 (a) Fatigue crack growth for specimens A and B at different stress levels and **(b)** optical images of specimen B with RTA 800°C at differet fatigue cycles.

DISCUSSION

The experimental results indicated that the deposition and RTA parameters significantly influence the mechanical properties of PECVD silicon nitrides. The results obtained by this particular experiment shows that a lower deposition temperature and a less chamber pressure could achieve a less residual stress and a better interfacial strength. Both of them are highly desirable from mechanical design perspective. On the other hand, deposition at lower temperatures and a less chamber pressure would possibly results in much lower deposition rate and incomplete chemical reactions to leave more residues in the deposited film. Meanwhile, by nanoindentation tests, it can be seen that an additional RTA could also improve the interfacial strength.

The fracture toughness of nitride films is another important factor for structural longevity concern since it directly impact the fast fracture and fatigue crack propagation during service.

Fracture toughness is usually characterized by indentation method. That is, by indenting a specimen with known force and measuring the generated crack size, it is possible to obtain the toughness by using the model proposed by Lawn and Marshall [5] or Zhang et al [6]. It is found that the fracture toughness was also improved with RTA. However, since the associated residual tensile stress was also increased. Eventually, the capability to resist fracture was actually decreased. This fact can also be seen in the fatigue test results shown in Figure 5. The fatigue growth of as-deposited specimen B was very low. However, after a 800°C RTA, the fatigue crack growth became more readable since the stress intensity factor increased after RTA. This implies that although the adhesion and toughness can be improved by RTA, the accompanied increase in tensile stress could actually make the structure less reliable in fracture and fatigue of nitride films. As a result, it is important to consider the all possible consequences in fabrication, mechanical reliability, and electrical functionality for a process design optimization.

CONCLUSIONS

In this work, the effects of different processing parameters and after-deposition RTA on the mechanical properties of PECVD silicon nitride films were characterized. Through curvature measurement, indentation testing, and fatigue test, it was found that deposition at a lower temperature with a smaller chamber pressure could result in a better quality in residual stress and interfacial strength between nitride and silicon. On the other hand, the residual stress and fracture were also changed after RTA. In particular, it is important to point out that although RTA could effectively enhance the adhesion and fracture toughness, the increase in residual tensile stress could actually reduce its capability to resist fast fracture and fatigue crack growth and it is important to consider the all possible consequences in fabrication, mechanical reliability, and electrical functionality for a process design optimization.

ACKNOWLEDGMENTS

This work is supported by National Science Council under the contracts NSC 96-2628-E-006-006-MY3 and NSC 97-2221-E006-1582-MY3.

REFERENCES

[1] I.K. Lin, P.H. Wu, K.S. Ou , K.S. Chen and X. Zhang, *Mater. Res. Soc. Symp. Proc.* **Volume 1222**, Warrendale, PA, 2010),1222-DD02-20(2010)
[2] G. G. Stoney , *Proceedings of the Royal Society*, **82**, 172(1909).
[3] P.H. Wu , Master Thesis , Mechancal Eng. Dept. National Cheng-Kung University, Taiwan, 2009
[4] W. C. Oliver and G. M. Pharr, *J. Mater. Res.*, **19**, 3(2004).
[5] H. Huang, K. Winchester, A. Suvorova, B. Lawn, Y. Liu, X. Hu, J. Dell, and L. Faraone, *Material Sci. Eng. A*, **435-436**, 453-459(2006)
[6] D. B. Marshall and B. R. Lawn, *ASTM Spec. Tech. Publ.*, **889**, 26(1986).
[7] T.Y. Zhang, L.Q Chen and R. Fu, *Acta Materialia*,**47**,3869(1999)
[8] T. L. Anderson, *Fracture Mechanics Fundamentals and Applications*,(CRC Press, 1991)

Mater. Res. Soc. Symp. Proc. Vol. 1299 © 2011 Materials Research Society
DOI: 10.1557/opl.2011.65

Effect of Phosphorus Doping on the Young's Modulus and Stress of Polysilicon Thin Films

Elena Bassiachvili[1] and Patricia Nieva[1]
[1]University of Waterloo, Waterloo, Ontario, Canada.

ABSTRACT

On-chip MEMS (Micro Electromechanical Systems) characterization devices have been used to extract the Young's modulus and average stress of polysilicon doped with phosphorus using thermal diffusion from a spin-on-dopant source. A customized fabrication process was developed and the devices were fabricated and tested. Resonant and static deformation tests were performed using microbridges. Information gathered from these experiments was combined to extract the Young's modulus and residual stress of the thin film. Several doping concentrations, from undoped to 2.99×10^{20} phosphorus atoms/cm^3 (4.148×10^{-4} Ω/cm), have been studied and it has been concluded that the Young's modulus of phosphorus doped polysilicon with a chemical phosphorus concentration of 1.96×10^{20} atoms/cm^3 (4.572×10^{-4} Ω/cm) increases by approximately 50GPa and the average stress of polysilicon with a phosphorus concentration of 2.99×10^{20} atoms/cm^3 (4.148×10^{-4} Ω/cm) becomes more tensile by approximately 63 MPa relative to undoped specimens.

INTRODUCTION

Material properties play a major role in the design process of MEMS devices and fabrication steps have long been used to alter material properties. Silicon, specifically polysilicon, has been the staple material used in MEMS device fabrication as well as electronics manufacturing. Phosphorus doping is often used to increase the conductivity of polysilicon. However, in addition to the electrical effects, these dopants also produce mechanical changes by causing changes in residual stress[1], Young's modulus[2], coefficient of thermal expansion[3] as well as other mechanical parameters. At the macro level, the characterization of material properties is a standard procedure. It is well known that the thin-film material properties vary from bulk material properties and one cannot use a thick sample to determine thin-film material properties [4]. Different approaches, such as nanoindentation [5], AFM bending [6], micro-tensile test [7] and others, have been taken to characterize the Young's modulus and stress of thin films. In this work, on-chip MEMS devices were fabricated and tested in order to extract the mechanical properties from the devices' responses. On-chip devices are, by their very nature, made directly from the material to be tested. This, along with the simplicity of the on-chip device, gives very direct access to the material properties of interest. Furthermore, an array of devices of various sizes located very close to each other can be used to cross-verify the extracted values.

EXPERIMENTAL DETAIL

Fabrication

On-chip MEMS characterization test devices, namely microbridges (i.e., clamped-clamped beams), were made using a customized fabrication process. Thin films with different dopant concentrations were produced and then patterned using the process flow shown in Figure 1 and described below.

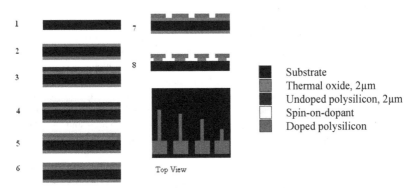

Figure 1. Process flow

The fabrication process starts at step 1 with a 100mm, 1-20 Ω/cm, <100> orientation silicon wafer, in step 2, a 2μm wet thermal oxide is grown on the wafer at just above 1000°C. The oxide is used to mechanically connect the polysilicon to the substrate as well as to electrically isolate it. In step 3, a 2μm layer of undoped LPCVD polysilicon is deposited at 625°C on the polished side of the wafer. In Step 4, a Filmtronics P512[1] spin-on-dopant is spun on and hard-baked at 200°C for 15 minutes. In step 5, the wafer is placed in a diffusion furnace at 1000°C for a pre-determined amount of time in an atmosphere of 75% N_2 and 25% O_2. The amount of diffusion time for each processed wafer differs and this produces samples with varying average dopant concentrations. In order to keep the thermal budget of all the wafers the same, and avoid possible changes of mechanical properties due to this parameter, each sample is pre-annealed prior to the dopant application. The pre-annealing time is calculated such that each wafer spends the same amount of time in the furnace. The resulting phosposilicate glass film, formed from the layer of spin-on dopant once the liquid evaporates, is removed in step 6 using a 1% HF dip and the doped polysilicon is patterned using reactive ion etching in step 7. The wafer is then diced and, in step 8, the structures are released by removing the oxide using 49% HF. The HF etches the oxide isotropically, each device is connected to an anchor plate large enough for a significant amount of the oxide to remain underneath and connect the device to the substrate. Figure 2 shows a

---

[1] A manufacturer of various spin-on films, http://www.filmtronics.com/

picture of an array of fabricated and released bridges of different lengths and widths. The dimensions were chosen to produce resonant frequencies within the range of controllable excitation of the piezoelectric shaker and appropriate ranges of other testing equipment used. The lengths varied from 150 μm to 850 μm and the widths varied from 2 μm to 30 μm. The narrowest properly-fabricated beams were used for testing in order to approximate an ideal beam as closely as possible.

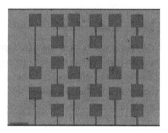

Figure 2. Micrograph of fabricated clamped-clamped bridges

Thin films with different dopant concentrations were produced by varying the amount of time a wafer spent in the furnace with the spin-on dopant with the resultant characteristics listed in Table 1. The values in Table 1 were obtained from several averaged 4-point-probe readings taken over the surface of the wafer. The average resistivity is calculated using the predicted junction depth from expected diffusion profiles, the average carrier concentration is further extracted from that [4]. The chemical concentration of phosphorus is interpolated from Mousty's experiments [8].

Table 1. Average resistivity of each wafer and the calculated average carrier and atomic concentration

Doping Time (hr)	Average Resistivity (Ω/cm)	Average Carrier Concentration (e$^-$/cm^3)	Average Chemical Phosphorus Concentration (atoms/cm^3)
0	1.83×10^4	0	0
0.5	2.68×10^{-3}	1.2×10^{19}	1.2×10^{19}
1	6.036×10^{-4}	9.1×10^{19}	1.06×10^{20}
1.5	5.376×10^{-4}	1.1×10^{20}	1.37×10^{20}
2	4.572×10^{-4}	1.4×10^{20}	1.96×10^{20}
2.5	4.148×10^{-4}	1.8×10^{20}	2.99×10^{20}

Experimental Setup

The fabricated MEMS test devices were tested at room temperature using resonant and static deformation tests. In the resonant test, a 10mm x 10mm piezoelectric shaker from CeramTec was used to produce out-of-plane excitation. An Agilent 33220A Function Waveform Generator was used to provide the electrical signal to the piezoelectric shaker and a Polytec OFV-551 vibrometer was used to measure the out-of-plane displacement of the microbridge. For the static tests, out-of-plane buckling and deformation of the microbridges post-release was observed using an optical profilometer and analyzed.

DISCUSSION

The resonant frequency of a clamped-clamped beam, f_0, with a Young's modulus E and average stress σ can be predicted analytically using Equation 1[9].

$$f_{0\ stress} = f_0 \sqrt{\left(1 + \frac{B\sigma L^2}{Et^2}\right)} \qquad \text{Equation 1}$$

Where B= 0.295. The experimental results can be matched to the analytical predictions to extract the required values. Since there are two parameters and only one equation, another experiment that is dependent on the Young's modulus and stress needs to be considered. The static out-of-plane buckling of microbridges can be used for this purpose. Once the microbridge beam buckles, the amount of out-of-plane deflection can be defined using Equation 2.

$$\sigma = \frac{E\pi^2}{12L^2}(3d^2 + 4t^2) \qquad \text{Equation 2}$$

Where t is the thickness of the film and d is the amount of deflection in the center of the beam. By combining the result from the resonant and buckling experiments and the equations presented above, the Young's modulus and average stress can be extracted for the available doping concentrations.

Figure 3 shows the frequency responses of five beams with the same dimensions and dopant concentration. Note the consistency of the resonant frequencies from sample to sample. The average resonant frequency was calculated using several samples for each dopant concentration and was then used.

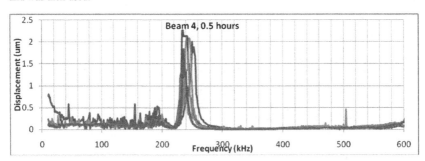

Figure 3. Examples of frequency signatures of five different samples of the same size and dopant concentration

Figure 4 shows the profilometer generated three-dimensional scan of three microbridges with the same dopant concentration but different lengths and widths. The out-of-plane buckling corresponded very closely with the beam length and decreased with higher dopant concentration. The as-deposited polysilicon film is concluded to have been initially highly compressive, due to the high out-of-plane deflection seen in Figure 4, and the introduction of the dopant atoms produces tensile stress which servers to counteract the intrinsic compressive stress.

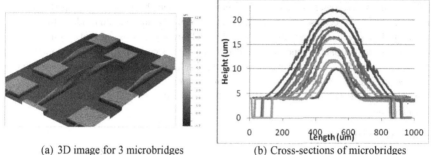

(a) 3D image for 3 microbridges   (b) Cross-sections of microbridges
Figure 4. Profilometer results for a dopant concentration of 1.96×10^{20} atoms/cm3

Figure 5 shows the resultant Young's modulus and stress values derived using the combination of the resonant and the static deformation tests and analytical models.

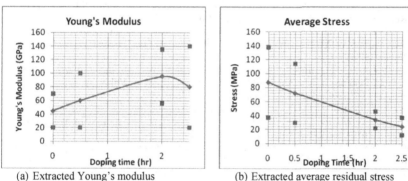

(a) Extracted Young's modulus   (b) Extracted average residual stress
Figure 5. Extracted mechanical parameters using the buckling and resonant test results for different doping times

The Young's modulus is found to generally increase with an increase in doping time, and hence the dopant concentration. Table 1 can be used to correlate the doping time to the average resistivity or chemical dopant concentration. The decrease in the Young's modulus for the 2.5 hour doping period could be attributed to saturation of the lattice by the dopant and the formation of a lower Young's modulus precipitate in the sample. The residual stress tends to become more tensile with higher doping times and dopant concentrations. Since phosphorus has a smaller atomic radius than silicon (P = 1.07Å, Si = 1.176Å) [10] its introduction into the lattice causes lattice contraction, so this result is expected.

CONCLUSIONS

Microbridges have been designed and fabricated using polysilicon thin films with different phosphorus concentrations. A resonant test and a static buckling test were performed and the results have been used to extract the effective Young's modulus and residual stress of the thin films. It has been concluded that the Young's modulus of phosphorus doped polysilicon with a chemical phosphorus concentration of 1.96×10^{20} atoms/cm^3 (4.572×10^{-4} Ω/cm) increases by approximately 50GPa and the average stress of polysilicon with a phosphorus concentration of 2.99×10^{20} atoms/cm^3 (4.148×10^{-4} Ω/cm) becomes more tensile by approximately 63 MPa. Due to high error ranges, only the trends, rather than the hard values should be taken into consideration.

ACKNOWLEDGMENTS

The authors would like to thank Natural Sciences and Engineering Research Council of Canada, CMC Microsystems and Ontario Centres of Excellence. Additionally, we would like to thank Madhavi Singaraju and Edward Xu for their help with fabrication.

REFERENCES

1. *Effect of phosphorus doping on stress in silicon and polycrystalline silicon.* **S.P. Murarka, T.F. Retajczyk Jr.** 4, April 1983, J. Appl. Phys., Vol. 54, pp. 2069-2072.
2. *Mechanical properties of phsophorus-doped polysilicon films.* **S.W. Lee, C.H. Cho, J.P. Kim, S.J. Park, S.W. Yi, D.I. Cho.** November 1998, Journal of the Korean Physical Society, Vol. 33, pp. 392-395.
3. *Stress and thermal expansion of boron-doped silicon membranes on silicon substrates.* **B.S. Berry, W.C. Pritchet.** 4, July/August 1991, J. Vac. Sci. Technol. A, Vol. 9, pp. 2231-2234.
4. **Sze, S.M.** *VLSI Technology.* New York : McGraw-Hill, 1988.
5. *Micromechanical and tribological characterization of doped single-crystal silicon and polysilicon films for microelectromechanical systems devices.* **B. Bhushan, X. Li.** 1, January 1997, Journal of Material Research, Vol. 12, pp. 54-63.
6. *Measurement of Young's modulus on microfabricated structures using a surface profiler.* **Y.C. Tai, R. Muller.** 1990. IEEE Xplore. pp. 147-152.
7. *Elastic properties and representative volume element of polycrystalline silicon for MEMS.* **Cho, S.W., Chasiotis, I.** 2007, Experimental Mechanics, Vol. 47, pp. 37-49
8. *Relationship between resistivity and phosphorus concentration in silicon.* **Mousty, F., Ostoja, P., Passari, L.** 10, October 1974, Journal of Applied Physics, Vol. 45, pp. 4576-4580.
9. *Measurement of Young's modulus and internal stress in silicon microresonators using a resonant frequency technique.* **Zhang, L.M., Uttamchandani, D., Culshaw, B., Dobson, P.** 1990, Measurement Science Technology, Vol. 1, pp. 1343-1346.
10. *Lattice parameter study of silicon uniformly doped with boron and phosphorus.* **Celotti, G., Nobili, D., Ostoja, P.** 1974, Journal of Materials Science, Vol. 9, pp. 821-828

Mater. Res. Soc. Symp. Proc. Vol. 1299 © 2011 Materials Research Society
DOI: 10.1557/opl.2011.66

Residual Stress in Sputtered Silicon Oxycarbide Thin Films

Ping Du[1], I-Kuan Lin[1,2], Yunfei Yan[1] and Xin Zhang[1]
[1] Department of Mechanical Engineering, Boston University, Boston, MA 02215, U.S.A.
[2] Global Science & Technology, Greenbelt, MD 20770, U.S.A.

ABSTRACT

Silicon carbide (SiC) has received increasing attention on the integration of micro-electro-mechanical system (MEMS) due to its excellent mechanical and chemical stability at elevated temperatures. However, the deposition process of SiC thin films tends to induce relative large residual stress. In this work, the relative low stress material silicon oxide was added into SiC by RF magnetron co-sputtering to form silicon oxycarbide (SiOC) composite films. The composition of the films was characterized by Energy dispersive X-ray (EDX) analysis. The Young's modulus and hardness of the films were measured by nanoindentation technique. The influence of oxygen/carbon ratio and rapid thermal annealing (RTA) temperature on the residual stress of the composite films was investigated by film-substrate curvature measurement using the Stoney's equation. By choosing the appropriate composition and post processing, a film with relative low residual stress could be obtained.

INTRODUCTION

Silicon-based ceramics (silicon oxide, silicon nitride, silicon carbide and their composites) are important structural and electronic materials for various applications in semiconductor and microelectromechanical systems (MEMS) [1]. Among these applications, mechanical properties such as modulus, hardness and residual stress play key roles to determine the device performances and structural reliability [2-4]. Typically, these ceramics can be fabricated by plasma-enhanced chemical vapor deposition (PECVD) [5] or magnetron sputtering [6] techniques. However, PECVD inevitably introduces a certain amount of hydrogen due to the precursors [7], therefore sputtering was used in this work. The thin film materials may have different properties from their bulk counterparts. Residual stresses, usually induced during the film formation, may have negative effects in thin-film processing: excessive compressive stress can result in film buckling and delamination, whereas excessive tensile stress may lead to film cracking [8]. In either case, residual stresses may damage the structures or reduce their service lifetime. Therefore, it is of vital importance to identify a reliable technique to minimize the residual stress in a controlled manner during the fabrication of MEMS.

In our previous study on co-sputtered silicon oxynitride films, we showed that the composition of oxygen and nitrogen had a strong influence on the mechanical properties of the composite films [9]. Both the Young's modulus and hardness increased with increasing nitrogen content in the composite films. In this work, we extend the material to silicon carbide, which is considered as a promising structural material with high strength at elevated temperatures [10, 11]. The distinct response of residual stresses in sputtered oxides and carbides will make it possible to tune the stress state in the film by properly controlling the composition of the composite films. In addition, annealing is reported to be a critical step for the residual stress. In this work, rapid thermal annealing (RTA) was employed to effectively reduce the residual stress with small thermal budget [12], and the effect of annealing temperature on the residual stress change was discussed with micro structure evolution.

EXPERIMENT

The oxycarbide films were deposited on silicon wafers by a Discovery 18 RF magnetron sputter (Denton Vacuum). The oxygen/carbon ratio was controlled by adjusting the RF powers applied to the pure (>99.9% purity) silicon dioxide and silicon carbide targets (Kurt J. Lesker), respectively. The flow rate of argon was 25 sccm (standard cubic centimeters per minute) to maintain a sputtering pressure at 4 mTorr during the deposition.

The composition of the SiOC films was characterized by A JSM-6100 scanning electron microscope (JEOL Ltd.) equipped with an EDX spectrometer (Oxford Instruments). The Young's modulus and hardness were measured by nanoindentation technique. A TI 900 Triboindenter (Hysitron Inc.) was used with a Berkovich indenter tip. The Young's modulus was obtained by fitting the unloading segment to an empirical power-law relation, and the hardness of the films can be determined from the maximum applied load divided by the maximum contact area [13, 14].

After deposition, the specimens were subjected to rapid thermal annealing (RTA) at temperatures between 400 – 800 °C using a RTP-600S system (Modular Process Technology Corp.) for 10 minutes. The curvature of the film-substrate system was measured by an Alpha-Step 500 Surface Profiler (KLA Tencor). The curvature was used to calculate the residual stress in the film by the Stoney's equation [15]

$$\sigma_f = \frac{E_s' t_s^2}{6 t_f}(\kappa - \kappa_0) \qquad (1)$$

. where σ is the residual stress, t is the thickness, κ is the curvature after deposition, κ_0 is the curvature of the substrate before deposition. E' is defined as the biaxial modulus and equals $E/(1-v)$, where E is Young's modulus and v is the Poisson's ratio. The subscript s stands for substrate and f for film.

RESULTS AND DISCUSSION

EDX analysis

The composition of oxycarbide film was controlled by adjusting the RF power applied to the oxide and carbide targets. Pure silicon oxide (SiO_x) and pure silicon carbide (SiC_y) specimens were deposited first. The deposition rate of oxide is roughly three times to that of carbide under the same RF power. Therefore the following three intermediate composites (SiO_xC_y1 to SiO_xC_y3) were obtained by simply varying the power applied to the oxide target. The sputtering conditions were summarized in Table I.

The films were investigated by EDX to determine the stoichiometric composition. Figure 1(a) shows the EDX spectra of all five films. The corresponding atomic concentration of Si, O and C in atomic concentration (at.%) are also summarized in Table I. The results indicate that with increasing carbon content in the SiOC films, the oxygen peaks continuously decrease while the carbon peaks continuously increase in the EDX spectra. The small amount of oxygen (5.76 at.%) in SiC_y could be due to the absorption of free oxygen in the SiC_y film surface.

Specimen	RF power (W)		Atomic concentration (at.%)		
	SiO₂	SiC	O	C	C/(C+O)
SiOₓ	300	0	85.38	0	0
SiOₓCᵧ1	50	300	63.23	29.38	31.72
SiOₓCᵧ2	100	300	53.87	38.52	41.69
SiOₓCᵧ3	200	300	42.53	49.56	53.82
SiCᵧ	0	300	5.76*	84.60	93.63

* Could due to the absorption of free oxygen in the film surface.

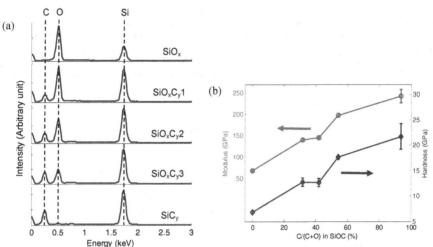

Figure 1. (a) EDX spectra of sputtered SiOC films. (b) The Young's modulus and hardness of as-deposited SiOC films with increasing carbon content [C/(C+O)].

Young's modulus and hardness

The Young's modulus and hardness were extracted from load-displacement data of nanoindentation. As shown in Figure 1(b), both the Young's modulus and hardness of SiOC films increase with increasing carbon content. For instance, the Young's modulus of SiOC films increased from 68.1 GPa to 242.9 GPa, and the hardness increased from 6.8 GPa to 21.7 GPa. This is in agreement with the literature that carbide is a much tougher material than the relative soft oxide. It also indicates that the existence of Si-C bonds embedded in Si-O matrix tended to enhance Young's modulus and hardness.

Residual stress

The residual stress in the composite films was calculated from the curvature of film-substrate system by using the Stoney equation. Before film deposition, the averaged initial curvature of the silicon wafer substrate was measured from the profiler as -2.05 x 10^{-2} m^{-1}. After deposition, the curvatures of SiOC films subjected to three RTA temperatures (400 °C, 600 °C and 800 °C) was measured and the residual stress was calculated with E'_s=179.4 GPa, v_s=0.28 and t_s=500 μm. The resulting residual stresses for different carbon contents under different annealing temperatures were plotted in Figure 2. For the as-deposited films (at room temperature of 23 °C), the residual stresses are generally compressive for all films. The development of compressive stress is associated with "atomic shot peening" mechanism [16]. During sputtering the sputtered atoms are ejected with energies in the order of 10 eV, about 100 times the energy of "evaporated" atoms. Moreover, the surface of the growing film is also submitted to bombardment by neutral argon atoms, namely ions that are neutralized at the cathode and backscattered with an energy less than, but in the same order of magnitude as, that of the accelerated ions. Both of these types of bombardment are prominent at low argon pressure. In such a film growing under bombardment the number of bonds under compression is higher than that under tension. Consequently such a process can put the whole film under a state of compression.

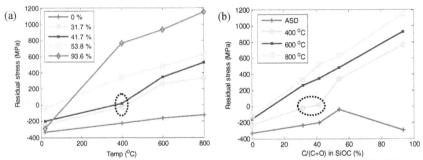

Figure 2. The influence of (a) annealing temperature and (b) carbon content [C/(C+O)] on the residual stresses of SiOC films. The films at minimum stress states are enclosed by dashed lines.

Because the sputtering is a relative low temperature process, the deposited films are generally amorphous [17]. The atoms deposited on the substrate lack sufficient kinetic energy to reach a thermodynamic equilibrium state and therefore result in a porous structure. Once they experience high temperature annealing, the increased mobility tends to drive the loosely ordered atoms to re-arrange to a relatively more ordered structure, and the density of the voids will be reduced [18]. Macroscopically the total volume decreases and this represents a net increase of tensile stress. This is clearly shown in Figure 2(a), where all five films experience an increase of residual stress. However, pure silicon carbide is more sensitive to the post-deposition thermal treatment. The residual stress changes from compressive (-298.7 MPa) at the as-deposited state to high tensile (768.3 MPa) after the 400 °C RTA treatment. It indicates that using only thermal treatment is not sufficient to control the residual stress of silicon carbide films.

As for the effect of carbon content, the residual stresses of all RTA treated films show monotonic increases from compressive to tensile with increasing carbon content (Figure 2(b)). Since they all experienced the same thermal treatment, the effect of extrinsic thermal stress should also be same. Therefore the difference is mainly due to the intrinsic stress, which is closely related to the microstructure in the films. The relative low melting point and soft (low hardness) SiO_x has low activation energy for self-diffusion of atoms and results in a low intrinsic stress. However, the high melting point and hard (high hardness) SiC_y has high activation energy for self-diffusion of atoms and results in a high intrinsic stress [19].

The minimum stress states of -21.16 MPa and 20.43 MPa are occurred for the SiOC films with carbon content of 31.7 % and 41.7 % after 400 °C RTA treatments, respectively. In other words, by properly controlling the composition between SiO_x and SiC_y, it is possible to minimize the residual stress of the composite materials.

CONCLUSIONS

Amorphous SiOC films with varied composition of oxygen and carbon content ranging from SiO_x to SiC_y were deposited by RF magnetron sputtering. The nanoindentation data show that both Young's modulus and hardness of SiOC films increases with increasing carbon content. In addition, the residual stress highly depends on the carbon content. By choosing the appropriate composition and post processing, a film with relative low residual stress could be obtained. This flexibility provides more opportunities for SiOC composite films to be integrated into the MEMS. The methodology presented in this paper can be also applied to material characterization of similar thin films.

ACKNOWLEDGMENTS

This work is supported by National Science Foundation under Grant CMMI-0700688. The authors gratefully acknowledge the assistance from Professor Catherine Klapperich at Boston University for her support with the nanoindentation instrument, Anlee Krupp at Photonics Center for EDX analysis, Andrew Bangert and Tianchen Liu for residual stress measurements.

REFERENCES

1. S. A. Campbell, *The Science and Engineering of Microelectronic Fabrication*. (Oxford, 1996).
2. X. Zhang, T. Y. Zhang, M. Wong and Y. Zohar, *Sens. Actuator A-Phys.* **64**, 109-115 (1998).
3. Z. Cao, T. Y. Zhang and X. Zhang, *J. Appl. Phys.* **97**, 104909 (2005).
4. Z. Cao and X. Zhang, *Sensor. Actuat A-Phys.* **127**, 221-227 (2006).
5. H. Yoshihara, H. Mori and M. Kiuchi, *Thin Solid Films* **76**, 1-10 (1981).
6. L. Guzman, S. Tuccio, A. Miotello, N. Laidani, L. Calliari and D. C. Kothari, *Surf. Coat. Tech.* **66**, 458-463 (1994).
7. C. M. M. Denisse, K. Z. Troost, J. B. O. Elferink, F. Habraken, W. F. Vandeweg and M. Hendriks, *J. Appl. Phys.* **60**, 2536-2542 (1986).
8. X. Zhang, T. Y. Zhang, M. Wong and Y. Zohar, *J. Microelectromech. Syst.* **7**, 356-364 (1998).

9. Y. Liu, I. K. Lin and X. Zhang, *Mater. Sci. Eng. A-Struct. Mater. Prop. Microstruct. Process.* **489**, 294-301 (2008).
10. D. Choi, Ph.D. Thesis, Massachusetts Institute of Technology, 2004.
11. M. Mehregany, C. A. Zorman, N. Rajan and C. H. Wu, *Proc. IEEE* **86**, 1594-1610 (1998).
12. M. W. Putty, S.-C. Chang, R. T. Howe, A. L. Robinson and K. D. Wise, *Sens. Actuator* **20**, 143-151 (1989).
13. W. C. Oliver and G. M. Pharr, *J. Mater. Res.* **7**, 1564-1583 (1992).
14. W. C. Oliver and G. M. Pharr, *J. Mater. Res.* **19**, 3-20 (2004).
15. G. G. Stoney, *Proc. Roy. Soc. A-Math. Phy.* **82**, 172-175 (1909).
16. F. M. Dheurle and J. M. E. Harper, *Thin Solid Films* **171**, 81-92 (1989).
17. C. Ziebert, J. Ye, S. Ulrich, A. P. Prskalo and S. Schmauder, *J. Nanosci. Nanotechno.* **10**, 1-9 (2010).
18. K. S. Chen, X. Zhang and S. Y. Lin, *Thin Solid Films* **434**, 190-202 (2003).
19. J. A. Thornton and D. W. Hoffman, *Thin Solid Films* **171**, 5-31 (1989).

Mater. Res. Soc. Symp. Proc. Vol. 1299 © 2011 Materials Research Society
DOI: 10.1557/opl.2011.397

Solid Bridging during Pattern Collapse (Stiction) Studied on Silicon Nanoparticles

Daniel Peter[1], Michael Dalmer[1], Andriy Lotnyk[2], Lorenz Kienle[2], Alfred Lechner[3], and Wolfgang Bensch[4]

[1] Lam Research Corporation, SEZ Str. 1, 9500 Villach, Austria
[2] Institute for Material Science, Christian-Albrechts-Universität Kiel, Kaiserstr. 2, 24143 Kiel, Germany
[3] Microsystems Engineering, University of Applied Sciences Regensburg, Seybothstr. 2, 93049 Regensburg, Germany
[4] Inorganic Chemistry, Christian-Albrechts-Universität, Max-Eyth-Str. 2, 24118 Kiel, Germany

ABSTRACT

The high surface to volume ratio of nanoparticles allows a detailed experimental study of the surface phenomena associated with solid bridging. Besides bulk analyses, the local view on the structure and composition via HRTEM is particularly essential. 50 nm core shell particles consisting of a silicon (Si) core and a SiO_2 shell were used as model system to understand surface phenomena appearing for Si-based nanostructures. Evaporative drying from de-ionized water shows the most significant bridging effect based on SiO_2. There is only a localized deposition of oxides between the particles during the drying process and no overall oxidation. For the deposition material, silicates are the most likely candidates.

INTRODUCTION

The surface preparation of semiconductor wafers is an important part of the production process. One essential requirement for wet surface preparation (i.e., etching and cleaning) is to avoid any damage of the structured wafer surface. The most vulnerable patterns are high aspect ratio structures, which are typically capacitor over bitline, shallow trench insulator (STI), or future FinFET structures. The largest mechanical forces in a wet clean process are generally attributed to physical force assisted cleaning steps (e.g., megasonics, spray systems) [1] and the surface tension forces in the drying step [2].

The distinctive damage from surface tension forces is the connection of the former free ends of the structures after drying (Figure 1). The fixed end can either be ruptured or elastically bent . The latter case is important for micro-electromechanical systems (MEMS) [3], which differ from microelectronic structures mainly by the higher aspect ratios and the larger size of micrometers versus nanometers.

For a permanent damage, a sufficiently large sticking force [3] has to keep the structures in place against the restoring elastic moment after the liquid, and therefore the surface tension force, is removed. On MEMS devices, the main adhesive force for hydrophilic surfaces is hydrogen bonding, and for hydrophobic surfaces, van der Waals forces (vdW) [4]. Additionally, after the collapse, residues (e.g., silica) can accumulate around the contact site and form a solid bond which is called solid bridging [4-7]. Corresponding adhesive energies are expected to be larger than hydrogen bonding and vdW forces; however, the determination of the characteristic energies is complicated by the inhomogenous deposition of the silica residues. The aspect ratios of microelectronic structures, such as STI, approach the aspect ratios of MEMS devices, and therefore elastic deformation is becoming more likely. Additionally, the solid bridging may be enhanced by the increased chemical reactivity of the nanostructured surface [8] of microelectronic structures

The solid bridging phenomenon was studied with Si nanoparticles in de-ionized water which were subsequently removed from the liquid by drying. Analyses were performed in the dispersion as well as with the dry powder.

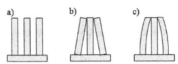

Figure 1. Pattern collapse of line structures: a) no collapse, b) rupture, c) bending.

EXPERIMENTAL DETAILS

Spherical silicon nanoparticles with a nominal diameter of 50 nm (Nanostructured & Amorphous Materials Inc.) were used in all the experiments in order to have a comparable length scale to STI structures and therefore similar nano-scale effects. Stable dispersions were created with de-ionized water (18.2 MΩ) using 10 min of ultrasonic sound (Allpax Palssonic). The concentration of the Si particles was adjusted to the dispersion limit (0.3 g/L) where no sedimentation occurs.

Two drying techniques were employed: first, evaporation drying with air flow at ~50°C (D (Table I)) and second, freeze-drying using liquid nitrogen (C). The latter was used in order to prevent oxidation during the drying step. Additionally, particles were aged in air for several months (B) and used as received (A).

Table I. Experimental conditions of the characterized nanoparticles.

Name	Aging condition	Drying
A (reference)	None	none
B	Air	none
C	dispersion (DIW)	freeze-drying
D	dispersion (DIW)	evaporation drying

The sizes of the dispersed particles were determined by light scattering techniques, and the presence of silicates was examined with electro-spray ionization mass spectroscopy (ESI-MS). The sizes and overall compositions of the particles were characterized by X-ray powder diffraction (XRD) (PANalytical X'pert PRO) with the Debye-Scherrer equation (using Cu-$K_{\alpha 1}$ and $K_{\alpha 2}$) [9] and transmission electron microscopy (TEM) (Tecnai F30 G^2 ST), respectively. In order to increase the accuracy of the XRD experiments, an internal standard of aluminum oxide (α-phase) was used. The intensity data were quantified via Rietveld refinement (FullProf [10]). For the study of the surface of the particles, X-ray photo electron spectroscopy (XPS) was applied, and solid state ^{29}Si NMR with magic angle spinning (MAS) was employed for the determination of the chemical composition of the silicon oxide. The local accumulation of SiO$_2$ was visualized by energy-filtered TEM (EFTEM) and high-resolution TEM (HRTEM).

RESULTS & DISCUSSION

Dispersion

The dispersions of the Si nanoparticles were stable in de-ionized water for more than 6 months without showing sedimentation. Light scattering measurements showed a zeta potential of about -30 mV; thus electrostatic interaction between the particles probably stabilizes the dispersion. The broad distribution of particle sizes as determined by the optical methods indicated a large increase of the average value of ~500 nm compared to the original sizes of ~50 nm. Sizes larger than 200 nm are due to agglomerates of the original particles as evidenced by TEM analyses. Nevertheless, as the optical methods are more sensitive to the larger particles/agglomerates, these methods could not be used for tracing the change of the small particles.

ESI-MS has been used to study the formation of silicates in water [11, 12] without disrupting the Si–O bonds. In the dispersion liquid, silicates were also measured and a trend of condensation to larger silicate species was observed from the initially formed mono-silicates similar to the trend observed by Pelster et al [13].

Chemical Analysis of The Dry Particles

The oxide shell on the silicon particles was evaluated by XPS (with the excitation energy of 1487 eV) and ^{29}Si MAS-NMR . A shift of the energies by 7 eV was observed in the XPS spectra (using O 1s signal) and attributed to sample charging. For the calculation of the oxide thickness on the surface of the silicon particles [14], the intensities of the Si 2p peaks were fitted by two Gaussian peaks (Figure 2). Only negligible signal intensity for Si^0 could be found for particles B, which corresponds to an oxide thickness in excess of 10 nm. All other particles, including the reference particles, showed an oxide thickness of ~5 nm. Hence no net oxide growth could be observed for the particles treated with de-ionized water. The calculated curve for particles A (Figure 2) using the two Gaussian functions closely follows the experimental values except for a small part in between the peaks .

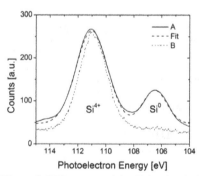

Figure 2. XPS spectra of the Si 2p signal of particles A and B.

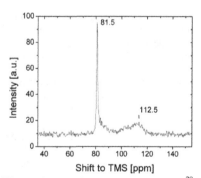

Figure 3. Particles A measured with ^{29}Si MAS-NMR.

This difference indicates the presence of a small amount of suboxides. The NMR data (Figure 3) did not show significant contribution from suboxides. The broad peak between 108 to 114 ppm is assigned to SiO_2 showing some variation of the Si-O-Si angles of neighboring SiO_4 tetrahedra [15]. Combining the NMR and XPS data indicates that oxide present on the particles is SiO_2. Additionally, the 81 ppm peak [16] in the NMR data points to crystalline silicon for the core of the particles which was confirmed by TEM measurements (Figure 8).

Size and Structure Analysis of the Dried Particles

X-ray powder diffraction and TEM micrographs were used supplementarily for the measurement of the sizes and the characterization of the microstructures. The former established the average size over billions of particles and the latter localized measurements of individual particles. The XRD data (Figure 4) were evaluated with Rietveld refinement because of the small differences in the reflection profiles and hence particle sizes (Figure 5).

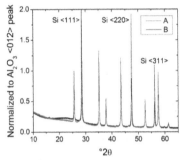

Figure 4. XRD pattern of the Si nanoparticles with Al_2O_3 (α-phase) as standard where only the silicon reflections are labeled.

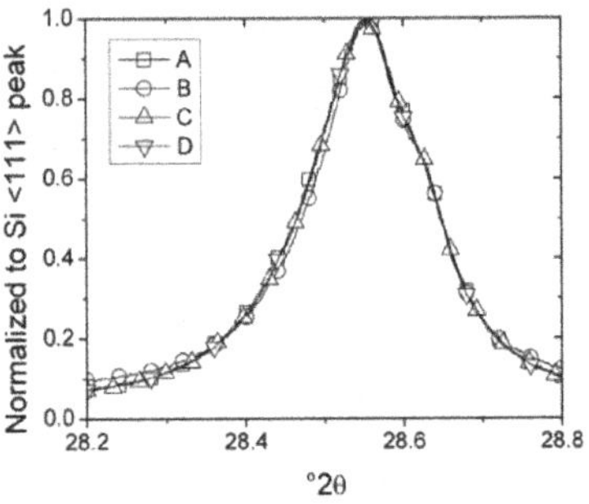

Figure 5. (111) reflection of Si measured by XRD for the various particle conditions. Only every 20th point is marked for better visibility.

Significant amorphous SiO_2 was found only on the particles B with XRD (Figure 4), which is consistent with the XPS observations on these particles. Due to the long time scale of several months, the oxidation in air appears more significant than the oxidation in de-ionized water. The particles B provided the proof that the other particles (A, C, D) are not at their native oxide limit, as well as that the XRD measurement is sensitive enough to show significant amorphous oxide growth. According to the XRD data, the average sizes of the particles were found to increase by 4 nm for particles C and 5 nm for particles D. However, the size difference cannot be due to the homogenous growth of SiO_2 on the surface of the particles, as indicated by the XPS data.

Table II. Sizes of the particles calculated from XRD data applying the Scherrer equation.

Particles	A	C	D
Size [nm]	60	64	65

The average sizes of the particles estimated from XRD data were confirmed by TEM bright field micrographs, which showed mostly spherical particles, generally in the range between

20 nm to 100 nm (Figure 6). The particles exhibit a core–shell microstructure. SiO$_2$ forms the shell as highlighted by the bright intensity at the particle edges in the EFTEM image of Figure 7. The Si in the core is crystalline as indicated by the lattice fringes in the HRTEM micrograph (Figure 8), which is consistent with the NMR data.

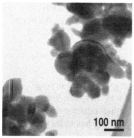

Figure 6. TEM micrograph of agglomerates of particles on the sample grid.

Figure 7. EFTEM image of oxygen distribution in the Si particles.

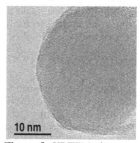

Figure 8. HRTEM showing the crystalline nature of the core Si particles.

Local Oxide Growth

The dependence of the sizes of the particles and the aging conditions suggests dissolution of the smallest particles and of the asperities into silicates, with the subsequent deposition as SiO$_2$ onto the larger particles. As there was no homogenous oxide growth on the surface of the particles, only a localized oxide growth was possible. This was confirmed by TEM measurements on the edge of the particle agglomerates (Figure 9-11), which showed SiO$_2$ accumulation between the particles, i.e., solid bridging. The size and the number of the solid bridges increase depending on the aging conditions. For the particles A, solid bridges were found rarely and with small sizes (<5 nm) which increased only in size for particles C (5-8 nm), whereas for particles D, their amount was large and in the 10 nm size range.

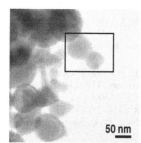

Figure 9. TEM bright field of particles D with solid bridging.

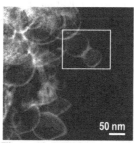

Figure 10. EFTEM image of oxygen distribution in the Si particles shown in Figure 9.

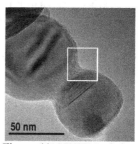

Figure 11. HRTEM of the box in Figures 9 and 10 with a solid bridge.

As the location of the preferential deposition is between the particles, the XPS as a surface-sensitive technique cannot detect the additional SiO_2, which explains the similar oxide thicknesses for the aged particles. However, the overall increase of amorphous oxide is very limited because no significant amorphous oxide was observed in the XRD patterns. Significant solid bridging was found for particles D, with limited bridges for particles C. Thus, the dominant step for solid bridging is the drying step. Freeze drying prevents an accumulation of SiO_2 between the particles probably because the ice cannot transport the silicates to the area between the particles.

CONCLUSIONS

Solid bridging between Si nanoparticles was found especially after evaporative drying processes as determined by TEM examinations. The bridging material was SiO_2 that was deposited during the drying process, most likely from silicates. For a semiconductor drying process where pattern collapse of nano-scale structures occurs, solid bridging has to be expected.

ACKNOWLEDGMENTS

The author would like to thank Dirk Meyer for the ESI-MS measurements. For the XPS measurements, the authors owe special thanks to Dr. V. Zaporojtchenko. The NMR data are courtesy of Prof. J. Senker (University of Bayreuth).

REFERENCES

1. C. De Marco, K.Wostyn, T. Bearda, K.-I. Sano, K. Kenis, T. Janssens, L. H. A. Leunissen, A. Eitoku, and P.W. Mertens, *ECS Transactions*, **11** (2) 87 (2007).
2. H. Cao, P. Nealey, and W. Domke, *J. Vac. Sci. Technol. B*, **18** (6), 3303 (2000).
3. C. H. Mastrangelo, *J. Microelectromech. S.*, **2** (1), 33 (1993).
4. K. Komvopoulos, *Wear*, **200**, 305 (1996).
5. R. L. Alley, G. J. Cuan, R. T. Howe, and K. Komvopoulos, Solid-State Sensor and Actuator Workshop, 5th Technical Digest,. (IEEE, Hilton Head Island, SC, 1992) pp. 202-207.
6. R. Maboudian and R. T. Howe, *J. Vac. Sci. Technol. B*, **15** (1), 1 1997.
7. R. Legtenberg, H. A. C. Tilmans, J. Elders, and M. Elwenspoek, *Sensors and Actuators A*, **43**, 230 (1994).
8. J. V. Stark, D. G. Park, I. Lagadic, and K. J. Klabune, *Chem. Mater.*, **8** 1904 (1996)
9. A. L. Patterson, *Physical Review*, **56**, 978 (1939).
10. J. Rodriguez-Carvajal, Fullprof.2k, Version 4.6c – Mar 2002, *Physica B*, **55**, 192 (1993).
11. S. A. Pelster, R. Kalamajka, W. Schrader, and F. Schüth, *Angew. Chem.*, **119**, 2349 (2007).
12. B. B. Schaack, W. Schrader, and F. Schüth, *Angew. Chem.*, **120**, 9232 (2008).
13. S. A. Pelster, W. Schrader, and F. Schüth, *J. Am. Chem. Soc.*, **128**, 4310 (2006)
14. F. J. Himpsel, F. R. McFeely, A. Taleb-Ibrahimi, J. A. Yarmoff, and G. Hollinger, *Physical Review B*, **38** (9), 6084 (1988).
15. E. Dupree and R. F. Pettifer, *Nature*, **308** (5), 523 (1984).
16. W.-L. Shao, J. Shinar, and B. C. Gerstein, *Phys. Rev. B*, **41** (3), 9491 (1990).

Mater. Res. Soc. Symp. Proc. Vol. 1299 © 2011 Materials Research Society
DOI: 10.1557/opl.2011.533

Fabrication and Characterization of Two Compliant Electrical Contacts for MEMS:

Gallium Microdroplets and Carbon Nanotube Turfs

Y. Kim, A. Qiu, J. A. Reid, R.D. Johnson, and D. F. Bahr
Mechanical and Materials Engineering, Washington State University, Pullman WA USA

ABSTRACT

Because of their high mechanical compliance and electrical properties, the idea of using Ga and CNTs for micro electrical relay contacts has been investigated to minimize damage from switching and make good electrical contacts. Ga was electroplated into droplets on the order of 50 μm in radius on single crystal Si to create a contact for a switch that can be annealed to recover its original electrical properties after mechanical damage. CNTs were grown on Si substrates, coated with a thin Au layer, and transferred to other Si or Kapton substrates through thermocompression bonding. In the case of the Ga contact, repeated switching led to an increase in the resistance, but the resistance recovered after a thermal reflow process at 120 °C. Longer term and larger area contacts were used to measure the contact behavior under switching conditions of up to 200 A/cm^2. At moderate cycling conditions (on the order of 200 cycles) the adhesion began to significantly degrade the switch. The oxidation behavior of the Ga droplets was characterized for thermal reflow, suggesting a passivating 30 nm oxide forms at 100 °C. The oxide formed by the Ga is thin and fragile as demonstrated by its use in a switch. The Ga droplets were examined with electrical contact resistance nanoindentation and the loads at fracture and the onset of electrical contact were identified. CNT turfs were also tested for making patterned electrical contacts; turfs of lateral dimensions similar to the Ga droplets were tested using electrical resistance testing during nanoindentation and as macroscopic contacts, and shown to be able to carry similar current densities. The results will be compared between the two systems, and benefits and challenges of each will be highlighted for creating compliant electrical switches and contacts.

INTRODUCTION

Micro electro mechanical systems (MEMS) cover a broad base of systems which use processing developed for microelectronics processing to fabricate devices with both electrical and mechanical functionality [1]. There are various types of micromechanical switches in MEMS [2, 3, 4]. These switches are useful in applications for a wide operating temperature ranges, radiation insensitivity, and usually have a high on-off impedance ratio. Ideally when the switch is closed it will exhibit good electrical and thermal. Almost all micromechanical switches have been designed with solid-to-solid contacts [5], and traditionally suffer from problems such as contact bounce, noise, high contact resistance, slow rise times, and a short operational lifetime due to mechanical wear and tear [6]. These problems may be solved by using compliant materials because they can exhibit a fast signal rise time, form high contact area at low applied loads, high thermal conductivity, and in some cases a low contact resistance [6, 7, 8]. Gallium (Ga) and carbon nanotube arrays (CNTs) are representative compliant materials. Gallium (Ga) proves to be a promising liquid metal above 29 °C, as its high boiling point (approximately 2400 °C) allows for Ga to be used in high current and high temperature applications. Small confined droplets of Ga have been studied on the nanoscale and have mechanical responses that imply a non-zero shear modulus even when liquid [9]. CNT turfs have been used for a wide variety of applications as an electrically conducting medium [10]. CNT turfs, arrays of CNTs in nominally vertical orientations, have been shown to be more compliant than carbon nanofiber arrays [11]. Therefore, this current study will focus on Ga and CNT turfs as compliant materials for electrical contacts in MEMS.

EXPERIMENT

Figure 1 shows the schematic process flow and fabrication of the Ga samples. The samples were fabricated from single side polished 3 inch diameter (100) boron doped silicon (Si) wafers. Silicon dioxide (SiO_2) was grown as a thin (120-180 nm) oxide layer using wet oxidation after boron doping. After patterning with photoresist, the oxidation layer of the sample was etched in Buffered Oxide Etch (BOE). A 10nm thick titanium / tungsten (Ti/W) adhesion layer was DC sputtered on the wafer, and 100nm thick W was then sputtered on the wafer. After photolithography the metal layers were etched in H_2O_2. Finally, the wafer was cleaned with with acetone, isopropyl alcohol, and deionized (DI) water and diced.

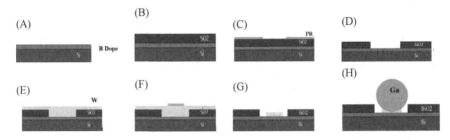

Figure 1. The process flow and the fabricated structure for Ga. (A) Boron doping, (B) Oxidation, (C) Patterning the oxide surface with the photoresist (PR), (D) Etching the patterned oxide layer in BOE, (E) sputtering Ti/W and W, (F) Patterning the W surface with PR, (G) Etching the W layer in H_2O_2, (H) Electroplating Ga on the sample.

The electrolyte consisted of gallium chloride and 140 ml solution of deionized (DI) water and hydrochloric acid (HCl). The pH of the solution was kept between 1.5 and 2.5, adjusted by adding HCl, and the concentration of gallium was $2.5M$. The conditions for the most uniform deposition used a current density, a cathode current, and a cathode potential of $1A/cm^2$, between 0.06A and 0.09A, and -0.6 to -0.7 V respectively, and the temperature was kept between 40°-50°C. A platinum wire and the sample were placed in a glass beaker as a H_2 evolution anode and cathode respectively [12]. Ga selectively plated on the W surfaces with only small amounts depositing on the B doped Si.

Figure 2 schematically shows the process flow and the fabricated structures for CNT turfs samples. Patterned CNT turfs were prepared with photolithography to create turfs 200 μm diameter with a height of approximately 20 μm. A sol gel of TEOS with solute Fe was deposited on clean silicon wafers in a manner described previously [13]. The individual tubes on these turfs are on the order of 15 nm in diameter and are multi-walled with 3-6 walls, previously identified using transmission electron microscopy. The turfs were grown in a tube furnace using chemical vapor deposition at a pressure of 100 mbar and temperature of 973 K, with a gas flow of 385sccm hydrogen and 25sccm ethylene. Growth time was varied between 30 and 60 minutes, and the heights of the turfs were between 21 and 23 μm. The tops of the CNT turfs were coated with a 5 nm Ti/W adhesion layer and 300 nm Au film, and then the CNT turfs were transferred a Si wafer coated with a 5 nm Ti/W and 300 nm Au film through thermocompression bonding at 150 °C for 2 hours [14].

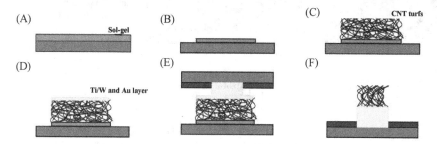

Figure 2. The process flow and the fabricated structure for CNT turfs. (A) Depositing sol gel on Si wafer, (B) Patterned sol gel, (C) Growth of the CNT turfs, (D) Sputtering Ti/W and Au on the tops of the CNT turfs, (E) Thermocompression bonding with patterned Au film, (F) Transferred CNT turfs to the patterned Au film. Procedures described in [14].

The thermal reflow of Ga was carried out in a furnace and ambient atmosphere. Reflow test samples were put in a pre-heated furnace. After the desired time was reached, the samples were cooled by switching off the furnace. To analyze the oxidation of the Ga structures that may occur during operation, TGA was carried out using a Rheometrics STA 650. Samples were placed in open crucibles in oxygen gas and were heated at 10 °C/minute to determine if the oxide was passivating.

To apply a voltage to the test sample, a W probe was affixed to the bonding pad, and the Ga droplets were contacted with other W probe (for short term) and with an aluminum sphere tip (for long term) in a compression test fixture to measure both load and displacement during contact. The current values were measured with a digital oscilloscope by monitoring the voltage across a resistor in parallel. A load of 0.5 N was applied for long term switch. Electrical contact resistance (ECR) nanoindentation in a Hysitron Triboindenter was used with a boron doped diamond tip to monitor the resistance during the initial stages of contact between a probe and the Ga droplets or CNT turfs. In this case the voltage and current were monitored using a Keithley® model 2602 dual channel system SourceMeter. The ECR nanoindentater was used for the effect of the current density as a function of force. A voltage of 2V was applied for ECR testing.

RESULTS and DISCUSSION

Figure 3 shows scanning electron microscope (SEM) images of electroplated and annealed Ga droplet, and the patterned and transferred CNT turfs. In general, before thermal reflow the surface of the deposited Ga droplet was rough, but after thermal reflow the surface was relatively smooth, and the droplet became relatively spherical in shape (Figure 3A).

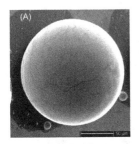

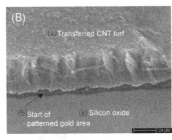

Figure 3. SEM pictures of (A) electroplated gallium (Ga) droplets after thermal reflow at 100°C for 10 min and (B) patterned and transferred CNT turfs.

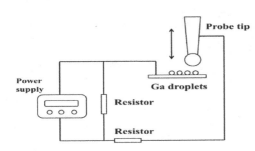

Figure 4. Schematic of the bulk electrical test set up.

The schematic of the test set up is shown in figure 4. Figure 5 shows the typical electrical performance of the Ga droplet switch at a constant voltage (3V DC) as a function of time with a W tip. Repeated switching on multiple droplets showed that the second contact always exhibited a lower resistance contact due to flattened the contact surface (leading to a larger contact area) after the first contact. However, the steady state resistance increases with the third and further contacts, as shown in figure 5 (A). It is assumed that this is because of plastic deformation and sub-surface defect generation. Subsequent thermal reflow (figure 5 (B)) "heals" the switch back to the initial performance level, suggesting damage to the switch is removed during reflow, previously identified using SEM [15].

Longer term and larger area contacts were used to measure the contact behavior under more realistic switching conditions. The contact was cycled at 0.4 Hz in depth control to apply a nominal load of 0.5 N (large enough to cause plastic deformation), and the applied load and current through the switch were monitored. This magnitude of contact leads to approximately four Ga droplets coming into contact with the aluminum probe tip, a total contact area of approximately 50 µm in radius, and a load of approximately 125 mN per droplet. The resulting current density of the switch for this test was approximately 200 A/cm². Each cycle produced an open and closed cycle until cycle 182. At cycle 182 the switch remained closed while the load was removed, suggesting material had transferred to the aluminum contact and fibers of Ga were bridging the gap.

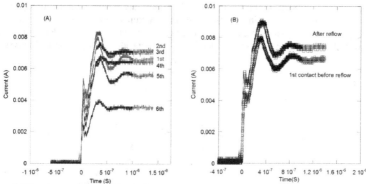

Figure 5. The short term switching behaviors of the Ga droplet: (A) First series of electrical switching tests showing degradation with continued switching. (B) Electrical performance after thermal reflow at 120 °C, where performance now exceeds the initial conditions.

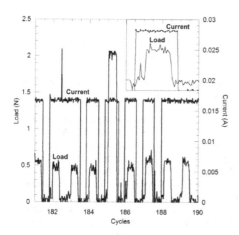

Figure 6. Cyclic mechanical switching at 0.5 V using Ga switch in contact with aluminum sphere at nomimal contact load of 0.5 N, inset shows cycle 186.

    Figure 6 shows the load and current during switching at cycle 182. The maximum load increased at cycle 185 in figure 6. The inset in figure 6 shows the switch turns on when the load begins to increase and remains on while the load decreases. In case of the solid-to-solid contacts, this phenomenon is exhibited when the contact surface has a clean and flat surface. According to Dickrell et al., a conductive contamination layer can develop with repeated contact during long term switching [16]. They implied that the contamination layer can be breached by added force.

    To examine the oxide thickness, after the switch was annealed at elevated temperatures TGA was performed. TGA analysis of a Ga sample was completed in oxygen. The sample was heated to 115 °C over the course of one hour (similar to the annealing conditions to recover the performance of the switch). TGA weight gain curves show gallium forms a protective oxide at temperatures above ambient. Using the number of droplets and weight gain, the oxide thickness after annealing in oxygen is estimated to be 30 nm. A thin oxide supported on a compliant substrate suggests that pressures in excess of those needed to cause permanent deformation of the substrate would cause fracture any passivation layer and make electrical contact in this system.

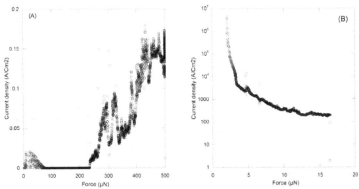

Figure 7. The current density values as function of force for (A) Ga and (B) CNT turfs using the ECR nanoindentation test.

Figure 7 shows the current density of a Ga droplet and CNT turfs as a function of force by ECR nanoindentation testing. The area of indentation was assumed to be equivalent to the area function of the diamond tip (i.e. to first order this ignores pile up and sink in) The current density value of Ga at a maximum force of 500 μN (3000 nm depth) was approximately 0.175 A/cm^2, whereas that of CNT turfs at 16 μN (200 nm depth) was approximately 190 A/cm^2. For the Ga droplet, as the force increased, the current density also increased except for the force range of 70 to 220 μN in figure 7 (A). The drop in current density in Figure 7(A) may be due to contamination of the oxide layer on the surface of the Ga droplet, and the subsequent rise in current would likely be attributed to the oxide film breaking. On the other hand, in the case of the CNT turfs, Figure 7(B), the current density decreased with an increase in the force because the resistance was increasing with increased contact area. We suspect this is due to significant adhesion at the beginning of the indentation with an underestimate of the actual contact area at the start of the indent; this will be the subject of a future study. The current density carried by the CNT turfs was significantly higher than that of the Ga droplet.

CONCLUSIONS

The suitability of a Ga micro-droplets and CNT turfs for compliant electrical contacts in MEMS was investigated by characterizing the electrical performance and nanoindentation behavior of the structures. Smooth Ga droplets can be formed using electroplating and thermal reflow processing at moderate temperatures. The electrical resistance of the Ga droplets increased due to mechanical deformation, but returned to the initial value after thermal reflow. The oxide on the Ga droplet formed by this thermal reflow is easily fractured with applied pressure to make electrical contact. Larger contacts could be run for over 100 cycles with no significant increase in resistance. In the larger contacts, the current density of Ga was approximately 200 A/cm². The current density measured using ECR nanoindentation increased as the applied force increased, whereas that of CNT turfs decreases as the force increased. However, the values of CNT turfs were significantly higher than that of Ga. Therefore, electroplated Ga micro droplets and CNT turfs may be used for electrical switches and contacts because moderate damage that may occur during switching of Ga droplets can be eliminated with a moderate thermal reflow process, and the CNT turfs can carry high current density under a very small amount of applied force.

ACKNOWLEDGEMENTS

This work was supported in part (AQ and DFB) by the National Science Foundation under the NIRT program, grant CMMI-0856436.

REFERENCES

1. A. Witvrouw, H.A.C. Tilmans, and I. De Wolf, *Microelectron. Eng.* 76, 245 (2004).
2. K. Petersen, *Proc. of the IEEE* 70, 420 (1982).
3. S. Roy, and M. Mehregany, *IEEE MEMS Workshop*, 353 (January-February 1995).
4. E. Hashimoto, Y. Uenishi, and A. Watabe, *Proc. Int. Solid-State Sensor Actuator*, 361 (June 1995).
5. J. Kim, W. Shen, L. Latorre, and C.J. Kim, *Sensor Actuator* A 97-98, 672 (2002).
6. J. Simon, S. Saffer, and C.J. Kim, *IEEE MEMS'96 Proc.*, 515 (1996).
7. M.M.J. Treacy, T.W. Ebbesen, and J.M. Gibson, *Nature* 381, 678-680 (1996).
8. P. Kim, L. Shi, A. Majumdar, and P.L. McEuen, *Phys. Rev. Lett.* 87, 215502 (2001).
9. J. B. Huang, G. T. Fei, J. P. Shui, P. Cui, and Y. Z. Wang, *Phys. Status. Solidi.* A 194, 167 (2002).
10. F. Kreupl, A.P. Graham, G.S. Duesberg, W. Steinhögl, M. Liebau, E. Unger, and W. Hönlein, *Microelecron. Eng.* 64, 349-408 (2002).
11. M. Terrones, N. Grobert, J. Olivares, J.P. Zhang, H. Terrones, K. Kordatos, W.K. Hsu, J.P. Hare, P.D. Townsend, K. Prassides, A K. Cheetham, H.W. Kroto, and D.R.M. Walton, *Nature* 388, 52-55 (1997).
12. S. Sundararajan and T.R. Bhat, *J. Less-common Met.*, 11, 360-364 (1966).
13. C.M. McCarter, R.F. Richards, S. Mesarovic, C.D. Richards, D.F. Bahr, D. McClain, J. Jiao, *J. Mater. Sci.*, 41, 7872-7878 (2006).
14. R.D. Johnson, D.F. Bahr, C.D. Richards, R.F. Richards, D. McClain, J. Green, and J. Jiao, *Nanotechnology*, 20, 065703, 6 (2009).
15. Y. Kim and D.F. Bahr, *Mater. Res. Soc.* Proc. 1139, (Boston, MA, 2008) GG03-05.
16. D.J. Dickrell, and M.T. Dugger, *IEEE Trans. Compon. Packag. Technol.* 30, 75 (2007).

AUTHOR INDEX

SUBJECT INDEX

Printed in the United States
By Bookmasters